URBAN
REGENERATION
IN THE UK

URBAN
REGENERATION
IN THE UK

PHIL JONES &
JAMES EVANS

Los Angeles • London • New Delhi • Singapore

© Phil Jones and James Evans 2008

First published 2008

SAGE Publications Ltd
1 Oliver's Yard
55 City Road
London EC1Y 1SP

SAGE Publications Inc.
2455 Teller Road
Thousand Oaks, California 91320

SAGE Publications India Pvt Ltd
B 1/I 1 Mohan Cooperative Industrial Area
Mathura Road, Post Bag 7
New Delhi 110 044

SAGE Publications Asia-Pacific Pte Ltd
33 Pekin Street #02-01
Far East Square
Singapore 048763

Library of Congress Control Number 2008922027

British Library Cataloguing in Publication data

A catalogue record for this book is available from the British Library

ISBN 978-1-4129-3490-9
ISBN 978-1-4129-3491-6 (pbk)

Typeset by C&M Digitals (P) Ltd., Chennai, India
Printed by The Cromwell Press Ltd, Trowbridge, Wiltshire
Printed on paper from sustainable resources

Contents

Acronyms

ABI	Area Based Initiative – refers to policy schemes tied to specific areas, as opposed to block grants given to local authorities for general purposes.
ASC	Academy for Sustainable Communities – a body administered by CLG with a remit to foster a culture of skills within the regeneration sector, although it does not itself engage in training.
AWM	Advantage West Midlands – RDA for the West Midlands.
BCC	Birmingham City Council – local authority for Birmingham.
BID	Business Improvement District – a locally-based initiative where businesses and property owners pay a voluntary additional tax to improve the environment of their local area.
BRE	Building Research Establishment – a government agency conducting and coordinating research on construction technologies.
BREEAM	Building Research Establishment Environmental Assessment Method – a measure of the performance of developments against certain indicators of environmental sustainability. EcoHomes is a domestic version of BREEAM.
CABE	Commission for Architecture and the Built Environment – statutory body set up by DCMS and ODPM to promote high-quality architecture and planning.
CBD	Central Business District
CIQ	Cultural Industries Quarter – district of Sheffield's inner city designated as a hothouse for the cultural industries.
CLG	(Department for) Communities and Local Government – successor to the ODPM, it's the main government department for urban policy in England since 2006.
CPO	Compulsory purchase order – mechanism through which local authorities and other government bodies can acquire property against the wishes of the landowner.
CPP	Community Planning Partnership – Scottish agencies, successor to the SIPs, aiming to improve indicators of social inclusion in the top 15% most deprived neighbourhoods in Scotland.
CPRE	Campaign to Protect Rural England – charity and lobby group seeking to protect the interests of rural England (formerly the Campaign for the Preservation of Rural England).
DBERR	Department for Business, Enterprise and Regulatory Reform – successor to the DTI, 2007.

DCMS   Department for Culture, Media and Sport – central government department which replaced the Department of National Heritage, in 1997.

DEFRA   Department for Environment, Food and Rural Affairs – main government department with responsibility for environmental policy, 2001.

DETR   Department of the Environment, Transport and the Regions – precursor to ODPM, 1997–2001.

DoE   Department of the Environment – central government department 1970–97, subsequently merged into the DETR.

DTI   Department for Trade and Industry – responsible for administering the RDAs, replaced by DBERR in 2007.

ECoC   European Capital of Culture – formerly European City of Culture, this EU-funded scheme seeks to promote the cultural heritage of individual European cities. The award is made annually on a rotation basis to different member states.

EEDA   East of England Development Agency – RDA for the east of England.

EP   English Partnerships – executive agency reporting to CLG. It is a significant landowner with a remit to help foster regeneration schemes across the UK in collaboration with local authorities, RDAs and other bodies (e.g. Pathfinders).

ERCF   Estates Renewal Challenge Fund – allowed local authorities to transfer individual estates into the ownership of housing associations. A smaller scale version of LSVT, the scheme ran between 1995 and 2000.

ERDF   European Regional Development Fund – funds made available by the EU to help even out regional inequalities within member states.

ESF   European Social Fund – EU's structural programme responsible for increasing skills and employment opportunities.

ESRC   Economic and Social Research Council – the main body for funding social science research in the UK higher education sector.

EU   European Union – a supranational body of European states which cooperate over certain aspects of social, economic and environmental policy.

GHA   Glasgow Housing Association – the housing association which took control of Glasgow City Council's housing portfolio following stock transfer in 2002.

GLA   Greater London Authority – a post-1997 replacement for the defunct GLC, with elections for the Mayor taking place in 2000.

GLC   Greater London Council – a powerful local authority which operated across Greater London and was abolished by the Conservative government in 1986.

HIP   Housing Improvement Programme – during the 1970s and 1980s this was the mechanism through which local authorities were allocated permission by central government to spend money maintaining their stock of council houses.

ICT   Information Communication Technology – umbrella term for computing and telecommunications.

LAA	Local Area Agreement – an agreement between central government, the local authority and LSP as to what the priorities are for action to improve local areas against 'floor targets' for education, health and public safety.
LDA	London Development Agency – the RDA for London.
LDF	Local development framework – flexible planning document produced at the area scale by local authorities. The intention is that they should function in a similar fashion to a development masterplan.
LDDC	London Docklands Development Corporation – the Urban Development Corporation with responsibility for regenerating the area around what is now Canary Wharf.
LSC	Learning and Skills Council – responsible for planning and funding education and training in England for those not in the university sector.
LSP	Local Strategic Partnership – responsible for delivering the Neighbourhood Renewal national strategy. LSPs map directly on to local authority boundaries and administer the use of the Neighbourhood Renewal Fund (NRF) in the 88 most deprived local authority areas.
LSVT	Large Scale Voluntary Transfer – introduced under the Conservative government, this has been the main mechanism for transferring the ownership of local authority housing stock to the housing association sector.
NAO	National Audit Office – a parliamentary body responsible for auditing the work of government departments, executive agencies and other public bodies.
NDC	New Deal for Communities – establishes local organisations to tackle indicators of social deprivation in specifically targeted areas, with no remit for physical reconstruction.
NRF	Neighbourhood Renewal Fund – sets 'floor targets' for improving indicators of social deprivation in the 88 most deprived local authority areas. It is administered at the local level by the LSPs.
NRU	Neighbourhood Renewal Unit – established in 2001, it is now part of CLG and oversees the government's Neighbourhood Renewal Strategy, administering the NDC, NRF and the LSPs.
ODPM	Office of the Deputy Prime Minister – it was responsible for urban policy for England, 2001–06, and was replaced by the CLG.
PFI	Private Finance Initiative – a mechanism whereby the private sector builds and maintains a capital resource such as a school or a hospital and leases it back to the state for a fixed period, often 25 years, after which it reverts to state ownership.
PPG	Planning Policy Guidance – guidance notes issued to local authorities on a variety of planning-related topics. Most of these were phased out in 2004–06, though some remain in force, where the advice has not fundamentally changed.

PPP   Public Private Partnerships – partnership arrangements between the state and private enterprise to deliver a particular project.

PPS   Planning Policy Statements – successor to the Planning Policy Guidance (PPG) notes, phased in from 2004.

PSA   Public Service Agreement – introduced in 1998, these set targets for performance and value for money in public services.

Quango   Quasi-Autonomous Non/National Government Organisation – a term popular in the 1970s and 1980s to describe executive agencies funded by central government but operating at one remove from direct democratic accountability. In the regeneration sector the term was classically applied to the UDCs.

RDA   Regional development agencies – established in 1998–99 with a remit to foster regional economic development. Transferred from the DETR to the Department for Trade and Industry (DTI) in 2001. In the same year the RDAs were given responsibility for distributing the Single Programme ('Single Pot') funding that replaced SRB.

RPA   Regional planning authorities – operate in England. Unelected bodies which have overall responsibility for producing the Regional Spatial Strategy (RSS) and overseeing the operation of the RDAs.

RSA   Regional Selective Assistance – discretionary grants available to encourage firms to locate or expand in designated Assisted Areas.

RSL   Registered social landlord – a body responsible for building and operating social housing while operating in the private sector with or without central government grants from the Housing Corporation. Often used as an alternative phrase for housing associations.

RSS   Regional Spatial Strategy – overall plans for how land is to be developed within a region over a 15–20-year period. These superseded Regional Planning Guidance in 2004. Although drawn up by the RPAs, RSSs must be approved by the Secretary of State for Communities and Local Government.

SAP   Standard Assessment Procedure – a measure of a building's energy efficiency.

SCP   Sustainable Communities Plan – launched in 2003, this sets out the government's long-term programme for delivering sustainable communities throughout England.

SEEDA   South East England Development Agency – RDA for south east England.

SFIE   Selective Finance for Investment in England – funds new investment projects that lead to long-term improvements in productivity, skills and employment.

SIP   Social Inclusion Partnership – Scottish agencies with a remit similar to the LSPs, which attempted to coordinate the actions of other agencies operating in an area towards promoting social inclusion. These were phased out 2003–04.

SINC	Site of Importance for Nature Conservation – a national network of non-statutorily protected wildlife sites, generally administered by local authorities in partnership with nature conservation organisations.
SME	Small and medium-sized enterprises.
SPD	Single Programming Document – a strategy document that maps priorities at the regional level with the objectives of the ERDF.
SPG	Supplementary Planning Guidance – produced by local authorities to cover a range of issues around a particular site, these are legally binding material considerations in subsequent decisions on planning permission.
SRB	Single Regeneration Budget – major national funding programme for urban regeneration, 1994–2001. Replaced by the Single Programme administered by RDAs.
SuDS	Sustainable (Urban) Drainage Systems – umbrella term for a collection of technologies which attempt to slow, reduce and purify discharges of rainwater runoff.
TEC	Training and Enterprise Council – executive agencies which operated at the regional level in the 1980s and 1990s with responsibility for fostering enterprise culture and economic development.
UDC	Urban development corporation – 1980s bodies set up by central government to bypass local authorities and undertake specific localised projects levering in private capital, e.g. London Docklands Development Corporation.
URC	Urban regeneration company – pioneered in the late 1990s, it became a central part of central government policy in the 2000 Urban White Paper. Established to act as a coordinating body (with no significant resources of its own) to bring together local parties to produce development plans for an area/city.
UTF	Urban Task Force – body headed up by architect Richard Rogers which had a significant influence on early new Labour thinking on cities. It produced *Towards an Urban Renaissance* in 1999.
WCED	World Commission on Environment and Development – also known as the Brundtland Commission, its 1987 report produced one of the first definitions of sustainable development.
WEFO	Welsh European Funding Office – agency of the Welsh Assembly government responsible for managing applications for funds from the European Union.

Overview

This chapter outlines the importance of urban regeneration within the UK context, and defines urban regeneration as a field of study. It then reflects on both the sheer scale of change and the wider political context in which these changes have occurred. Finally, the scope and structure of the book are outlined, with some guidance on how it should be used.

Introduction

Over the last decade it has become hard to ignore the almost continual process of development and building that has characterised the inner areas of many cities and towns. Anyone living in or visiting a UK city will be familiar with building sites that seem to sprout shiny new buildings overnight and the sight of cranes dominating the skyline. Such is the scale of the urban regeneration agenda that vast swathes of the UK's towns and cities are being pulled down and built anew. This process is profoundly transforming our urban areas, both in terms of their appearance and the ways in which we live in them. More than this, contemporary urban regeneration offers an important chance to rectify the mistakes of the past and create attractive places where people want to live in the future. This book examines what urban regeneration is and the ways in which it is changing our cities and towns.

What is 'urban regeneration'?

Cities are never finished objects; land uses change, plots are redeveloped, the urban area itself expands and, occasionally, shrinks. Pressure to change land uses can come about for a number of reasons, whether it be changes in the economy, environment, or social need, or a combination of these. The large-scale process of adapting the existing built environment, with varying degrees of direction from the state, is today generally referred to in the UK as urban regeneration. Some of the core elements of regeneration have appeared in urban policy before, albeit with slightly different labels. In post-war Britain there was a discourse of reconstruction, not only addressing areas which had suffered the destruction of wartime bombing, but also demolishing the large areas of slum housing which had been thrown up during the nineteenth century to house a growing urban industrial workforce. Urban reconstruction was somewhat akin to urban renewal in the United States, wherein large parts of the inner cities were demolished and replaced with major new roads, state-sponsored mass housing and new pieces of urban infrastructure.

Urban regeneration is a somewhat newer phrase, which arose during the 1980s and carries with it a particular emphasis. The urban sociologist Rob Furbey has critically reviewed this phrase, reflecting that 'regeneration' in Latin means 'rebirth' and it thus carries with it a series of Judaeo-Christian associations of being born again. This notion was particularly appealing during the 1980s, when urban policy under Prime Minister Margaret Thatcher swung towards the **neoliberal** and was influenced by a very particular vision of Christianity, centred on the individual rather than the broader community. In a sense, therefore, regeneration – as opposed to mere 'redevelopment' – became akin to a moral crusade, rescuing not only the economy but also the soul of the nation. The phrase also functions as a biological metaphor, with run-down areas seen as sores or cancers requiring regeneration activity to heal the body of the city (Furbey, 1999).

One of the most significant figures in the early history of urban regeneration was Michael Heseltine, who served as Secretary of State at the Department of the Environment between 1979 and 1983. This was a crucial period in the history of British cities, partly because Heseltine drove through the right-to-buy legislation which allowed tenants to buy their council houses at substantial discounts – a part-privatisation of social housing that massively increased owner occupation. Perhaps more significantly, Heseltine also led the government's response to the 1981 riots in the deprived inner-city areas of Handsworth in Birmingham and Toxteth in Liverpool. Heseltine concluded that something dramatic needed to be done and there followed a series of policy interventions attempting to redevelop derelict and under-utilised sites, bringing economic activity and social change to deprived areas.

The Conservative approach during the 1980s doubtless had its flaws, but it set a trend for large-scale interventions reconfiguring the urban fabric of areas suffering from economic decline following the shift away from a manufacturing-led economy. From relatively modest beginnings in the 1980s, regeneration has become a tool applied in almost all urban areas in the UK, accelerating in the past decade in parallel to rapid growth in the property market. Regeneration developed as an **holistic** term for the economic, social and environmental transformation of run-down urban areas. In recent years, however, there has been an interesting shift with social and community policy partially hived off into a **discourse** of neighbourhood *renewal* (not to be confused with the meaning of urban renewal in the United States). Regeneration as a concept has been somewhat diluted as a result and although the policy rhetoric retains the language of an holistic approach, regeneration does seem to have retreated to having a much greater emphasis on interventions in the built form to stimulate economic growth. It is in this area, rather than in community policy, that this book finds its focus.

Table 1.1   Housebuilding completions in the UK (CLG, 2007a)

	England	Wales	Scotland	Northern Ireland	UK
1998/99	140,708	7,737	20,637	9,638	**178,720**
1999/2000	142,046	8,706	24,214	10,399	**185,365**
2000/01	133,255	8,333	23,465	11,668	**176,721**
2001/02	129,866	8,273	23,610	13,487	**175,236**
2002/03	137,739	8,310	23,361	14,415	**183,825**
2003/04	143,958	8,296	23,662	14,511	**190,427**
2004/05	155,893	8,492	26,408	15,768	**206,561**
2005/06	163,398	8,257	24,482	17,410	**213,547**

The scale of change

Urban regeneration policies have played a significant part in creating a boom in the construction sector. In absolute terms the amount of construction occurring in the UK has increased by over 25% between 1995 and 2005, and the sector now contributes 8% of the UK's total gross domestic product (DTI, 2006b). This growth in development has been concentrated in urban areas. Although only around 10% of the UK's land surface is urbanised, the percentage of total new residences built in urban areas has grown from under 50% in 1985 to over 65% in 2003 (Karadimitriou, 2005).

The frenzy of building in UK towns and cities is not simply a product of economic growth, but reflects broader demographic shifts within the UK population. People are living longer than ever before and at the other end of the age scale, people are waiting longer to have children, both of which mean a decrease in average household size which, combined with growing population, means that the number of households is increasing rapidly. As a result, the number of households in England alone is predicted to rise from just over 21 million in 2004 to nearly 26.5 million in 2029 with 70% of that increase taking the form of one-person households (CLG, 2007a).

As indicated by Table 1.1, the completion rate of new-build dwellings in the UK has been steadily increasing since 1998. The number of households in Scotland is projected to rise by an average of 14,800 a year from 2004 to 2024, well within the current levels of housebuilding (General Register Office for Scotland, 2007). There is a similar story in Northern Ireland, with a predicted average annual increase of 7,800 from 2001 to 2025 (Northern Ireland Statistics and Research Agency, 2007). Wales, on the other hand, has a projected average annual increase of households from 2003 to 2026 of 10,500 per year, significantly above the rate of new dwelling construction (National Statistics, 2006). The situation in England is even more acute, with an average annual increase of 217,400 households predicted to 2029, far below the level of housebuilding.

The gap between the increasing numbers of households and the rate of housebuilding in England is one part of a very complex story of rapid house price inflation. In the government's response to the Barker Review of Housing Supply (discussed in Chapter 2), there was a commitment made to reaching a target of 200,000 new dwellings built per year in England within a decade (HM Treasury, 2005). It was felt that without this increase, on top of the already substantial increases since 1998, housing would become increasingly unaffordable and this would damage economic performance, constraining labour mobility and business competitiveness, particularly in the south east. There are also significant social consequences to consider where certain socio-economic groups can simply no longer afford to live in some parts of the country.

That much of the predicted and actual increase takes the form of one-person households is one factor behind the boom of apartment development within UK cities. This kind of development fits into a broader discourse of central city transformation, which has seen city centres become increasingly fashionable places to live, work and play. While urban regeneration is about more than city-centre redevelopment – a point addressed directly in Chapter 7 – there is no doubt that the pace and scale of inner-city change has been rapid, with new apartments, retail and leisure developments producing major changes in the patterns of land use. Regeneration bosses in Liverpool, for example, recently complained to Google that they were overlooking the regeneration of the north (BBC, 2006). The web application Google Earth, which allows users to look at bird's eye views of the world, was using aerial photographs from the late 1990s and so did not show either Liverpool's Paradise Street redevelopment, a 17-hectare £900m retail and leisure project in the city centre, or the Arena and Conference Centre, a new 10,000-seat venue being built on the waterfront. Other new buildings, such as Manchester's City of Manchester Stadium, which hosted the 2002 Commonwealth Games and is now home to Manchester City FC, were shown as derelict land.

As well as showing the pace of change, this story highlights another key element of urban regeneration – the importance of changing a city's image. Liverpool's regeneration planning director emphasised this, claiming that 'The city centre has changed dramatically. ... It is important that the millions of people using Google Earth have access to the latest images showing the city's transformation.' Urban regeneration thus constitutes a physical *and* a symbolic transformation. Part and parcel of rebuilding a city is to reinvent the city for a new generation. In order to understand the reasons for this dual character of urban regeneration, it is necessary to consider briefly the wider context within which urban regeneration has come to the fore.

The context for contemporary regeneration

One of the challenges of studying urban regeneration is that it is not an isolated process. Cities are affected by wider economic, political and environmental factors. The fortunes of cities are tied to the fortunes of nations and, ultimately, the global economy. Over the course of the twentieth century cities in the western world suffered from the loss of traditional industries that were undercut by cheaper products from east Asia or withered by the decline in colonial power. Across Europe and north America, urban regeneration began as an attempt to ameliorate the negative effects of deindustrialisation and enable cities to attract new investment in the global economy. The goal of policy was to direct development and investment towards those areas in which it was most needed. Left to their own devices, developers would chose to locate developments on the cheapest land in areas with the highest demand. In the UK context this would result in pressure to relax constraints on greenfield development, particularly around London and the south east. At its heart, therefore, regeneration is a political strategy using a whole range of planning regulations and policies to encourage developers to invest in run-down and derelict urban areas.

The Conservative government in the late 1980s and early 1990s allowed a large number of out-of-town developments which, though economically successful in themselves, damaged the economies of central cities while at the same time increasing car dependency. The Conservatives started to address this through the 1990s, but it was the **New Labour** government which came to power in 1997 which really brought a sea-change in attitudes towards urban development. **Brownfield** sites, previously developed land within existing urban areas, became the key strategic target for meeting housing and development needs. This strategy was given formal expression in 2000 when Planning Policy Guidance Note 3 was released setting a target for local authorities to build 60% of new housing on brownfield sites. This 60% target gave a significant boost to the urban regeneration agenda by forcing local authorities and developers to look first to target sites within existing cities. Although estimates of total amounts of brownfield land are notoriously inaccurate, its distribution follows the geography of deindustrialisation and hence much of this land is located in urban areas. Derelict land is frequently considered an eyesore and its redevelopment is a critical element in regeneration, replacing an undesirable land use with high-quality housing. The definition of what comprises a brownfield site is drawn rather broadly, however, and can include some rather surprising types of land uses, not simply derelict industrial sites. This point will be returned to in Chapter 5.

The 60% brownfield target helped reinforce a tendency for developers to build flats rather than houses within cities in order to maximise the number of residential units which can be fitted on to a development site. This helps local authorities meet their housing targets, and enables developers to maximise the returns from the purchase of expensive inner-city land. As a result, the number of new flats being built each year has grown from 23,626 in 2000 to 56,823 in 2006, while the increasing re-use of vacant land has caused the proportion of new development in urban areas to overtake that on rural land (Aldrick and Wallop, 2007).

The wider context allows us to understand *where* and *why* urban regeneration takes place – on brownfield land in cities that are flagging economically and/or that require more housing in order to stimulate economic growth. Perhaps the more important question this book seeks to answer is *how* such developments are to be built. The government is committed to the principles of sustainable development, and the need to balance economic, social and environmental factors cuts across not only urban regeneration, but public and planning policy more widely. While the concept of sustainability is notoriously difficult to pin down, it implies a commitment to protecting the environment and ensuring equal access to social and environmental services as well as economic development. In the face of climate change the question of how to make cities more sustainable is growing in importance. The form that new development takes has a direct bearing on environmental factors, such as how much energy is consumed and how much waste is produced. Further, the design of cities is now also being seen as a key way in which to adapt cities to a changing climate. The idea of 'quality of life' has become common parlance, as the political agenda has subtly shifted towards creating environments in which people want to be. Related to these trends, urban regeneration has been caught up in the wider 'new urbanism' movement that emphasises high-quality design and well-planned spaces. Urban regeneration therefore not only acts as a vehicle for reinventing the economies and tarnished reputations of declining industrial cities, but simultaneously helps deliver the government's policy agenda on sustainability.

As a primarily political agenda, the practice of regeneration is framed by wider trends in British politics. The withering of public funding throughout the 1980s and 1990s means that the public sector now works in partnership with the private sector to make urban regeneration happen. Issues of how to form effective partnerships and how to fund urban regeneration are very real challenges, as massive regeneration projects can entail hundreds of partners and hundreds of millions of pounds of funding. In order to understand *how* regeneration takes place it is necessary to look at the different policies that have been used over time to facilitate this process, and the different ways in which partnerships have worked.

The scope and structure of this book

A book of this kind has a lot of ground to cover. Historically, our focus begins with the emergence of 'urban regeneration' as a serious policy domain in the early 1980s but concentrates on developments since the Labour government came to power in 1997. While various precursors to regeneration are mentioned where necessary, there is no space for a more general history of urban development. Similarly, while much regeneration practice involves drawing on successful ideas used in other countries, the focus here is on the UK. The UK is, of course, a country of many parts, with distinct legal-political substructures for England, Wales, Scotland and Northern Ireland. In terms of the case studies used to illustrate the discussion, examples have been drawn from these four regional blocks, although inevitably with greater weight given to England, being significantly larger in population than the other three combined. Given the scale of regeneration activity in the UK over the past quarter century, it would be impossible to mention every interesting scheme which has been undertaken. While there is some degree of regional balance in the case studies chosen, there are inevitably omissions, particularly in terms of the smaller towns, whose regeneration schemes have tended to receive less attention than those of the larger metropolitan areas.

Within the UK field of regeneration there is a vast amount of published materials and the proliferation of academic journals focusing on regeneration is a good barometer of scholarly interest in the topic. Indeed, as this book was being written the *Journal of Urban Regeneration and Renewal* was launched. But despite its importance, urban regeneration does not fit neatly into existing disciplinary and sub-disciplinary categories, not least because it spans social, economic and environmental dimensions. Regeneration is driven by applied practice, rather than academic research. As a result, research tends to be scattered across a variety of disciplines, from more obvious ones such as urban studies and planning, to regional studies, public policy, property development and engineering.

For the same reasons, regeneration involves a bewildering range of government departments and organisations, all of which release reports, papers and research within the field. Government policy changes rapidly, as do the responses from various stakeholders. Further, many of the key organisations frequently change their names, making it even harder to keep tabs on the sector. For example, the government departments responsible for environment and communities were reshuffled four times between 1996 and 2006. The fast pace of policy change within the sector may explain the lack of textbooks dealing with regeneration.

This book aims to contextualise the regeneration agenda and synthesise existing research in a systematic way to provide a reference text for this emergent field. It is aimed primarily at an academic audience, as there are an increasing number of university courses dealing with urban regeneration. The book is designed for third-year undergraduates, postgraduates and academics and will take the reader through the basic context of regeneration into state-of-the-art research. Accordingly, the topics have been chosen to reflect core themes from the academic literature, rather than to act as a practical guide on how to 'do' regeneration. So, for example, we have not chosen to cover the legal aspects of regeneration – while these aspects play a crucial role in work on the ground, the technical elements are probably of less interest to a general academic audience. We have also aimed to strike a balance between covering the wider context for regeneration, while retaining a focus on regeneration itself. Transport, for example, though playing a major role in where and how regeneration can be undertaken, is not covered separately. Similarly, while social issues are discussed to provide context for urban regeneration policies and case studies, these are not given chapters in their own right. A number of texts already cover these issues in more depth than is possible in a general review of regeneration activity and these are indicated where appropriate.

The book is organised into eight chapters, with each chapter covering a distinct aspect of regeneration with illustrative case studies used throughout. Chapter 2 deals with the policy framework, detailing the legislative context in which urban regeneration operates. A brief overview is given of post-industrial policy approaches during the 1980s and early 1990s, before moving on to critically review urban policy under the Labour governments from 1997 (including the urban renaissance agenda and the impact of devolution). Chapter 3 considers issues of governance, in order to understand the political processes through which urban regeneration is actually delivered. The notion of partnerships, which is central to contemporary regeneration, is critically analysed. Chapter 4 explores the strategies for economic growth that underpin urban regeneration through the idea of the 'competitive city'. The chapter identifies key funding streams and approaches to urban economic regeneration and examines their success.

Chapter 5 tackles the issue of sustainability, which has become a central concept in all discussions of contemporary regeneration. Key social and environmental policies are reviewed and different approaches to integrated planning are assessed. Chapter 6 considers the visual transformation of the cityscape, examining issues of design and cultural elements of regeneration. A number of key tensions are explored, surrounding architectural innovation and the retention of heritage and as well as questions of culture and identity. Chapter 7 charts the extension of the urban regeneration agenda beyond central cities, to suburban and ex-urban developments, and asks whether the central city regeneration model can be transposed on to the suburbs. It also explores regional-scale mega-regeneration projects through an examination of the Thames Gateway. Finally, Chapter 8 summarises and integrates the key themes that span each chapter, and explores the future direction of developments within the sector.

How to use this book

This book has a number of features that are intended to make it easier to use. Most importantly, it makes extensive use of case studies to demonstrate how concepts and policies work in practice. The case studies are primarily drawn from academic research and are used to think critically about the advantages and disadvantages of different ways of doing urban regeneration. The book aims to give detailed descriptions and explanations of how urban regeneration works, while also questioning dominant approaches. At the start of each chapter there is an overview, summarising the contents, arguments and overall structure of the chapter. At the end of each major section within chapters key arguments are summarised to enable the major points to be easily identified. Each chapter ends with an annotated reading list that highlights key academic texts for each of the concepts addressed, and reading about the wider ideas that frame regeneration. This list allows the reader to undertake further research in specific areas of interest.

In addition to an index, the book also has a glossary of academic terms that are used (highlighted in bold), and an annotated list of acronyms to help guide the reader through the 'alphabet soup' of multiple agencies and policies. While the book has been designed as a coherent whole, with key concepts and cases cross-referenced within the text, each chapter can also be read as a stand-alone learning aid. Due to the fast-paced nature of the regeneration sector, we have also included a 'keeping up to date' section after the final chapter, which lists key websites and news feeds for regeneration in the UK.

2 Policy Framework

Overview

This chapter details the legislative context in which urban regeneration operates.

- *Introduction: the road to 1997*: gives a brief overview of the development of urban policy prior to the election of the New Labour government at the end of the 1990s.
- *Contemporary English Policy*: explores the main policies and agencies responsible for urban regeneration within England.
- *Devolution*: explores the policy landscape developed by the devolved governments in Scotland, Wales and Northern Ireland.

Introduction: the road to 1997

It is impossible to discuss urban regeneration without looking at the policy context in which it operates. The details of different urban policies can be a rather dry subject, but it is of critical importance to the way in which actors in the regeneration process are able to operate. For a little over a quarter of a century now the political context for urban regeneration has been broadly **neoliberal**. The election of Margaret Thatcher in 1979 crystallised an emerging belief that the state could no longer be the primary actor in the redevelopment of cities. Instead, the philosophy was one of market forces guiding the private sector to invest, with the state intervening only as far as it created the conditions for the private sector to step in.

There was an important political context in which this neoliberal shift took place. The financial crises of the 1970s required swingeing cuts to public spending. The Conservatives took office in 1979 determined to further rein in the public sector, which was seen not only as inefficient, but also giving too much power to the labour unions. Many city councils were controlled by the Labour party, which at that time was fighting its own internal battle against 'militant' hard-left tendencies, played out in cities like Liverpool. **Thatcherite** urban policy was directed towards greatly reducing the power of these local authorities as part of a broader assault on the political left. This did not, however, automatically mean that the private sector entirely took over urban redevelopment, but rather that central government took much more control over spending at local level, sidelining those Labour councils.

Competitive bidding

As part of the Thatcherite reforms a new principle was introduced to determine the level of funding that central government gave to local authorities to undertake urban redevelopment. The maintenance and regeneration of local authority housing estates had been funded through the Housing Improvement Programme (HIP), which left local authorities free to determine where they spent resources allocated within a block grant. While HIP was retained, new competitive bidding regimes were introduced which required councils to put proposals together for redeveloping individual estates and areas. These proposals would be evaluated alongside proposals from other local authorities within the region and funding allocated to the projects deemed most 'deserving'.

Schemes like Estate Action resulted in large injections of cash for relatively small areas, resulting in a kind of 'grand slam' approach to redevelopment. Local authorities had an incentive to put their most deprived areas into these competitions to increase their chance of winning funds against less deprived estates within other local authority areas – a kind of ugliness contest. This actually helped certain very deprived areas as there had been a tendency among some local authorities to concentrate resources on less run-down areas where they felt the money would do more good. The problem was that these competitive schemes were funded by reducing the overall HIP allocation, which meant overall cuts in general maintenance. This resulted in considerable neglect and decline of areas which were not successful in the competitions, with local authorities not permitted by central government to divert revenue from other areas into maintenance.

The Estates Action scheme was primarily targeted at upgrading areas of run-down council housing, but the principle of area-based competitive bidding which it developed became the model for more general funding in what, from the mid-1980s, was beginning to be called urban 'regeneration'. The City Challenge scheme contained an element of physical renewal in areas of council housing, but had a broader remit to foster the economic redevelopment of the target area. Rebecca Fearnley (2000) has examined a City Challenge-funded scheme based in the Stratford area of Newham in east London, which was seen by the government as one of the most successful of these projects. Fearnley notes that the Stratford scheme, which ran from 1993 to 1998, had some significant successes, such as an overall increase in housing satisfaction as well as decreases in reported crime and fear of crime in the area. She argues, however, that the scheme focused on issues which were comparably easy to tackle, such as physical renewal of the housing stock. Indeed, in terms of economic regeneration, while much work was done increasing the employability of residents, the scheme was much less successful at actually attracting employers to the area to increase the number of jobs available.

One of Fearnley's overall criticisms of City Challenge was that it mostly worked through the existing structures of local service delivery – local authorities and schools – and was much less successful at bringing in and nurturing community-led organisations and projects. In more recent urban policy there has been a much greater emphasis on the need to successfully bring the community into the process. If there is one fundamental shift that came out of Conservative neoliberal policies, it has been the need to bring together multiple actors – community, private sector and various state agencies – in order to undertake regeneration, even if today the neoliberal rhetoric has been somewhat softened. To give a simple example, it is no use bringing new employers into an area if the schools are not producing students with the necessary skills to fill the jobs. This central idea of bringing partners into regeneration projects will be discussed in more detail in Chapter 3 where questions of governance are addressed.

New institutional structures

As well as setting the general parameters for how central government funding schemes now operate, the neoliberal approach of the 1980s and 1990s had major implications for how regeneration would be organised. The term used at the time, although it is not much heard now, was the central government **'quango'** (quasi-autonomous national/non-governmental organisation). These essentially were arms' length executive agencies which wielded considerable power, but answered only to the relevant minister, rather than having any direct democratic line of accountability. In terms of urban regeneration, perhaps the most important quangos set up under Thatcher were the urban development corporations (UDCs).

The UDCs were parachuted into chronically deprived urban areas to bypass local authorities and attempt to stimulate a process of physical and economic renewal. The first two, London Docklands and Merseyside, were established in 1981 and eventually 11 others were put in place. These were limited-life organisations and were all wound up by the mid-1990s, with the exception of Laganside Development Corporation in Belfast, which ran until early 2007. The London Docklands Development Corporation (LDDC) was probably the best known, investing heavily in new infrastructure projects to help lever in major new private office developments. In spite of the collapse of the office property market in the late 1980s, which briefly left Canary Wharf looking dangerously like a white elephant, there is little doubt that there has been a radical improvement in the physical infrastructure and economic activity in the area – which were, after all, the main aims of the UDCs.

The LDDC was finally wound up in 1998 and produced a series of publications examining its own achievements. Reviewing these studies, Florio and Brownhill (2000) note that the somewhat heroic accounts of dramatic changes to the area brush over the considerable tensions that the LDDC created. The primary problem was that it represented the *redevelopment* of the area, not its *regeneration* – existing socio-economic problems in the area were not helped by the creation of a shiny new office cluster. Indeed, the argument is that the developments actually increased social polarisation by creating islands of extreme wealth while leaving untouched large neighbouring populations suffering acute poverty. For all of this criticism, however, it is interesting that the UDC model has recently been revived in order to meet some of the needs of the 2003 Sustainable Communities Plan, discussed below.

The later UDCs were considerably less well funded and it was clear that the model was really too expensive to be more generally applicable. By the 1990s, there was a degree of pragmatism among the British political left that the neoliberal agenda was here to stay with Labour-controlled local authorities accepting that they had to work within these strictures. In turn, under John Major's premiership, there was a softening of the stance on local authorities and a rehabilitation of these bodies as partners in the regeneration process. With their powers greatly curtailed, there would be no return to councils being able to take on much of the process themselves, but unlike the UDCs, they not only had expertise in physical renewal but also in community issues such as education, health and social welfare.

Policy continuity

The year 1997 and the election of a ('New') Labour government is as much a watershed in British politics as the election of Margaret Thatcher's Conservatives in 1979, but this was not necessarily immediately obvious at the time. Committed to the Conservative's spending plans during that early period in order to reassure middle-class voters, there was no sudden abandonment of neoliberal policy principles. In the first few years, key Conservative policies were retained, in particular the Single Regeneration Budget and attempts to move local authority housing out of council control and into the housing association sector.

The Single Regeneration Budget (SRB) was introduced in England in 1994 and drew together a series of different funding strands, with the idea of reducing complexity in the system. Unlike projects based around the UDCs or funding schemes such as Estates Action, the SRB Challenge Fund was not exclusively targeted at areas of acute deprivation. In the first three rounds of SRB funding, spending in the 99 most deprived areas amounted to £122.50 per head. The remainder of the country was not forgotten, however, with £21.30 per head spent in the other 267 districts designated 'non-deprived' (Brennan et al., 1999: 2074). Indeed, having rolled together a number of different programmes, the types of project which received funding could vary enormously, which was a significant advantage for taking an **holistic** approach to tackling complex socio-economic-environmental problems in an area.

SRB was originally administered by the Government Offices for the Regions – essentially regional offshoots of the Department of the Environment – and, post-1997, the Department for Environment, Transport and the Regions (DETR). As such it remained something which was very much controlled by central government. In 2001 it was announced that SRB would cease running, although existing projects would continue to be funded. The administration of these projects was handed over to the newly established regional development agencies (RDAs), which were given the remit to operate the replacement for SRB, the so-called 'single pot' (see below).

The SRB was still essentially predicated on competitive bidding and there was no attempt to move away from this post-1997. More controversially than the continuation of SRB funding during the early years of **New Labour**, however, was the acceleration of the programme of Large Scale Voluntary Transfer (LSVT). In neoliberal terms, this was a logical extension of right-to-buy legislation, which had reduced the overall size of council housing stock by encouraging sitting tenants to purchase their homes at a significant discount. Under LSVT local authorities were encouraged to transfer the ownership and management of their remaining council housing to housing associations. These housing associations, though eligible for public sector grants and regulated by the public sector, are effectively private sector, non-profit bodies that can borrow private finance – their activity does not therefore show up in measures of public spending/borrowing.

Some councils had assumed that with Labour returned to power in 1997, the very tight restrictions on how much could be spent maintaining their housing stock would be eased and the transfer policy scrapped. Such assumptions were rapidly scotched. Transfers required a vote in favour from tenants and the primary attraction was that transfer would bring with it a significant injection of new funds – with the implicit threat that housing stock would continue to be neglected for lack of resources if left with the local authority. The rate of transfers, which had stayed below 50,000 housing units per year under the Conservatives, topped 100,000 a year between 2000 and 2002. Indeed, of 133 tenant ballots between 1999 and 2004, only 16 resulted in a rejection of the transfer proposals (Ginsberg, 2005).

LSVT and the fact that local authorities are simply no longer permitted to build new homes has significant implications for urban regeneration. Where housing stock has not been transferred, it is now almost impossible to undertake significant changes to the physical infrastructure of council housing areas. Birmingham, which rejected stock transfer in 2003, continues to struggle to find investment for run-down areas of council housing. Glasgow, on the other hand, voted in favour of transfer, making it much easier to work on strategic schemes of demolition and rebuilding such as that associated with the bid to host the 2014 Commonwealth Games in the East End area of the city.

Key points

- Under the Thatcher governments local authorities were partially bypassed as agents of urban redevelopment, with urban development corporations used to lever in non-state partners and finance.
- The principle of competitive bidding for central government grants has become a key element in resource allocation, creating the suspicion that this allows the central state to set local priorities.
- The election of a Labour government in 1997 saw the acceleration, rather than reversal, of these principles of partnership and competition.

Contemporary English policy

This section deals with the current situation in England. Broadly similar policies also apply in Wales, although these are filtered through the different governance structures which have been put in place since the partial devolution of powers to the Welsh Assembly in 1999 (see below). The Thatcherite emphasis on bringing a variety of partners into any regeneration process has been retained and extended in current policy and the implications of this in terms of governance are discussed in the next chapter. Where Thatcher's governments expanded the number and variety of arm's length agencies dealing with regeneration, Labour governments since 1997 have taken this flowering of state agencies to dizzying heights of complexity. It can, in fact, be quite difficult to keep up even with changes to the formal institutions, let alone the policies being operated. Indeed, the complexity of the bureaucratic process has led in part to the establishment of the Academy for Sustainable Communities (yet another executive agency!) dedicated to building capacity within local communities and other non-state actors trying to negotiate the minefield of contemporary regeneration policy – of which see below.

From holistic regeneration to sustainable communities

Some of these changes can in part be explained by the rise and fall of John Prescott, the Deputy Prime Minister from 1997 to 2007. Prescott was given overall responsibility for urban policy in 1997 and a new 'super Ministry' (the Department for Environment, Transport and the Regions, DETR) was set up, integrating the old Department of the Environment with some functions taken from other departments to give a much more holistic approach when it came to regeneration. Labour's first term saw a whole variety of exciting and innovative urban policies being floated; this was the era of the Urban Task Force led by internationally renowned architect Richard Rogers and the establishment of the Commission for Architecture and the Built Environment (CABE). During Labour's second term, the DETR was replaced by the Office of the Deputy Prime Minister (ODPM), which retained the local government and regional portfolios, but lost much of the environmental remit to the newly established Department for Environment, Food and Rural Affairs (DEFRA) and the economic growth remit to the Department for Trade and Industry (DTI). With Prescott forced to relinquish his remaining departmental responsibilities following a personal scandal in 2006, the Department for Communities and Local Government (CLG) was created. Regeneration policy lost both the coherence that the DETR structures had brought and also the political clout of the Deputy Prime Minister – becoming just another department competing for resources from the Treasury. Although CLG retains responsibility for planning policy, the change of name to 'communities' is quite significant as it de-emphasises the macho world of altering physical forms. Instead, the new name reminds us that the point of urban regeneration is not new buildings and townscapes, but rather that reforms to the physical environment are just one part of making life better for *people* – improving society and communities. Nonetheless, in the rough, tough world of struggles for power between different branches of the civil services, 'communities' does have a rather weak feel to it. As the initially holistic remit of this key government department was slowly eroded, there was a marked impact on the direction of urban policy during the Blair regime.

Urban Task Force, Urban White Paper

The Urban White Paper, *Our Towns and Cities: The Future* (DETR, 2000), staked out the shape of urban policy at an early point in the Labour administration. It was based in part on the report of a task force commissioned by John Prescott to look at urban policy. Their report, *Towards an Urban Renaissance* (Urban Task Force, 1999), reflected the optimism of the period and had a strong leaning towards the importance of high-quality design – unsurprising given the involvement of Richard Rogers. Praise was lavished on cities like Barcelona, combining high-density housing, high standards of urban design and vibrant cultural identity. The Urban Task Force report was not without its critics (detailed by Cooper, 2000) and while the emphasis on urban design did find its way into the subsequent White Paper, it was not top of the priority list. Instead, issues of local involvement in decision-making, an emphasis on partnership working and a reinvigoration of local and regional government were emphasised first.

Perhaps the most important thing that came out of both the Urban Task Force report and the subsequent White Paper was a clear commitment to sustainability being at the heart of urban policy. Sustainability as a *concept* will be discussed in more detail in Chapter 5, but given its centrality to contemporary policy it must be briefly mentioned here. As the White Paper argued:

> We also have to bring together economic, social and environmental measures in a coherent approach to enable people and places to achieve their economic potential; bring social justice and equality of opportunity; and create places where people want to live and work. These issues are interdependent and cannot be looked at in isolation. ... That is why moving towards more mixed and sustainable communities is important to many of our plans for improving the quality of urban life. (DETR, 2000)

Note that there was a close link made between communities being 'mixed' and therefore being 'sustainable'. Mixing is not only about demographics – income, age, family structure, ethnicity, etc. – but also about that live/work/play mix in the built form that the Urban Task Force stressed. This holistic notion of sustainability integrating economic, social and environmental concerns had a relatively coherent bureaucratic form under the DETR, but in the reorganisation after the 2001 election this integration was lost.

Planning policy and the Barker reviews

In spite of responsibilities for the different aspects of sustainability being split across departments, it remains key to the entire policy discourse – no subsequent initiative has been launched without making at least some reference to sustainability. This was reflected, for example, in the launch of the replacements to the Planning Policy Guidance (PPG) series, which set the framework within which local authorities operate planning policy. The new documents are called Planning Policy Statements (PPS) (see Box 2.1) and the first in the series, replacing the old PPG1 *General Policies and Principles* was, significantly, titled *Delivering Sustainable Development*. The essence of this document is the assertion that: 'Plans should be drawn up with community involvement and present a shared vision and strategy of how the area should develop to achieve more sustainable patterns of development' (ODPM, 2005b: 3). Again, note the interdependence of strong, involved communities and sustainability. While practice may not live up to these visions, the rhetoric highlights an important shift of mindset towards actively pursuing a sustainability agenda post-1997.

Box 2.1   Planning Policy Statements/Guidance

In revising the existing Planning Policy Guidance (PPG) notes to a new series of Planning Policy Statements (PPSs), the ODPM prioritised the most potentially contentious areas in order to reduce uncertainty in the planning system that was a result of the pending changes. As such, some of the less contentious PPGs remain in force, sitting alongside the newer PPSs. Perhaps the most contentious of all was *PPS 3, Housing*, which was published slightly later in order to integrate the findings of the Barker Review of Housing Supply.

Planning Policy Statement 1: Delivering Sustainable Development (February 2005)
Planning Policy Guidance 2: Green Belts (March 2001)
Planning Policy Statement 3: Housing (November 2006)
Planning Policy Guidance 4: Industrial, Commercial Development and Small Firms (November 1992)
Planning Policy Guidance 5: Simplified Planning Zones (November 1992)
Planning Policy Statement 6: Planning for Town Centres (March 2005)
Planning Policy Statement 7: Sustainable Development in Rural Areas (August 2004)
Planning Policy Guidance 8: Telecommunications (August 2001)
Planning Policy Statement 9: Biodiversity and Geological Conservation (August 2005)
Planning Policy Statement 10: Planning for Sustainable Waste Management (July 2005)
Planning Policy Statement 11: Regional Spatial Strategies (September 2004)
Planning Policy Statement 12: Local Development Frameworks (September 2004)
Planning Policy Guidance 13: Transport (March 2001)
Planning Policy Guidance 14: Development on Unstable Land (April 1990)
Planning Policy Guidance 15: Planning and the Historic Environment (September 1994)
Planning Policy Guidance 16: Archaeology and Planning (November 1990)
Planning Policy Guidance 17: Planning for Open Space, Sport and Recreation (July 2002)
Planning Policy Guidance 18: Enforcing Planning Control (December 1991)
Planning Policy Guidance 19: Outdoor Advertisement Control (March 1992)
Planning Policy Guidance 20: Coastal Planning (September 1992)
Planning Policy Guidance 21: Tourism (was cancelled in May 2006 and replaced by the *Good Practice Guide on Planning for Tourism*)
Planning Policy Statement 22: Renewable Energy (August 2004)
Planning Policy Statement 23: Planning and Pollution Control (November 2004)
Planning Policy Guidance 24: Planning and Noise (September 1994)
Planning Policy Statement 25: Development and Flood Risk (December 2006)

The new policy statements came in the aftermath of the Planning and Compulsory Purchase Act, 2004 which placed sustainability of communities and environments – as well as the economy – at the heart of the planning system. The 2004 Act introduced a number of major reforms with the intention of streamlining the planning process. New regional spatial strategies were introduced, replacing the old regional planning guidance but, crucially, having statutory force. At the local level, older plans, such as the unitary development plans, were replaced by local development documents which have to work in accordance with the regional spatial strategy. The intention was to give developers more clarity and certainty about the process and to try to cut down the amount of time plans languished within the system.

This agenda of streamlining the planning process has been driven in part by the Barker reviews. Work on the first of these reports began in 2003 when Kate Barker, an economist and member of the Monetary Policy Committee, was asked by the Chancellor and the Deputy Prime Minister to produce a review of housing supply in the UK. When this review was commissioned, UK house prices had been rising steeply for a number of years and the Treasury was concerned that this was causing the economy to overheat, while the ODPM was concerned about affordability.

The first Barker report, *Delivering Stability: Securing Our Future Housing Needs* (Barker, 2004) argued that housing supply was not being mapped on to demand. Allocations of housing land were previously allocated by local authorities based on population projections. The concern was that where areas had low demand, too much housing land might be released for development, while high-demand areas might see local authorities refusing further planning permissions in a given development cycle where their existing demographic targets had been met. Barker proposed that allocations of housing land should be more closely related to the market price of land.

This was quite a controversial move as it was seen as further encouraging the growth of the south east of England, where there is clearly a high market demand for development land. Nonetheless, the review fed directly into the drawing up of *Planning Policy Statement 3: Housing* (PPS 3) which was released at the end of 2006. This is quite an interesting turn as it means that the Treasury is now taking a very direct role in shaping planning policy – resulting in a decidedly market-driven slant. There are a lot of positive reforms in PPS 3, particularly about the need to ensure high-quality design, the need to preserve/restore biodiversity and secure mixed communities through the provision of socially rented/affordable housing. Even so, lobby groups such as the Campaign to Protect Rural England (CPRE) fear that the need to forecast market needs up to 20 years ahead of time is potentially problematic. If demand is not as high as forecast, the CPRE argue, housebuilders will simply build at low densities to fill up the 'surplus' land (CPRE Oxfordshire, 2006).

The housing review which heavily influenced PPS3 was followed up by the *Barker Review of Land Use Planning* (Barker, 2006). This was in the same vein as the housing review in that it called for a more market-determined view on organising the release of land for more general development. Indeed, there was also a view that in certain circumstances some reconfiguration of the green belt surrounding urban areas might be appropriate. The obvious critique of this is that if not carefully managed, it could lead to a return to the boom of out-of-town shopping centres that occurred in the late 1980s. There is, after all, a clear market for such developments, although it would work against more general policy aims of revitalising urban centres and reducing reliance on car-based transport. The review also called for significant changes to the planning process in an attempt to give developers a clearer sense of what was required and thus speed up the processing of applications, with a slow planning process seen by the Treasury as a major brake on economic development. Even with this major Treasury involvement, it is interesting that the review is heavily couched in the language of sustainability, not, as would have been the case in the 1960s for example, on the need to promote 'growth'.

Regional policy

When the DETR was broken up in 2001, the new ministry, though retaining responsibility for the regions, lost control of regional economic policy. The regional development agencies (RDAs), which had only been established in 1998, were passed over to the Department for Trade and Industry (DTI). The loss of the RDAs was significant because it meant that the Ministry with responsibility for urban regeneration lost direct control over one of its major funding streams as the Single Regeneration Budget was scrapped and replaced with the RDA-administered Single Programme. The concern is that these bodies, with their statutory obligation to further economic development in the regions, should see regeneration merely as part of an economic agenda, rather than as an holistic process of which economic development is one part. Intriguingly, the money RDAs have to spend mostly comes from CLG but they have much less influence over how that money is spent, with the Department for Business, Enterprise & Regulatory Reform (DBERR, the successor to the DTI) remaining in overall control and the RDAs having considerable freedom in determining how they choose to spend the resource allocated to them (Greenhalgh and Shaw, 2003).

The original intention was that the RDAs would be matched to newly established regional assemblies, but these were to have very limited powers – nothing like what was given to the Welsh Assembly or the Scottish Parliament. With the exception of the London Assembly, none of the English Assemblies are directly elected. A referendum in the north east saw the proposal for a directly elected assembly resoundingly rejected in 2004. This has left an odd situation where the bodies established to give democratic oversight to the actions of the unelected RDAs are themselves without democratic mandate. These bodies are, however, charged with drawing up the regional spatial strategies. These cover transport, housing, economic development, the environment, tourism and regeneration, establishing at regional level the locations, size and priority of development – setting, for example, the amount and location of land to be released for housing development. The failure of the regional assembly policy following the 2004 north east referendum has, therefore, left a significant democratic deficit in these crucial areas of regional planning.

Urban regeneration companies and urban development corporations

The urban regeneration companies (URCs) were considered a major policy instrument for delivering the aims of the Urban White Paper (DETR, 2000). Based on three pilot URCs established in 1999, there are now 21 in England, one in Wales, five in Scotland and one in Northern Ireland. Unlike the old urban development corporations, they are not set up with the assumption that the local authority has failed and therefore needs to be bypassed. Essentially these are strategic partnerships funded by CLG (sometimes alongside English Partnerships) with the relevant regional development agency and local authority.

The idea is that the URC should set out a masterplan for the regeneration of a specific area. Public sector partners can then use this to prioritise the redevelopment of key infrastructure. This, in turn, can attract private capital to realise the rest of the plan. URCs themselves do not have significant resources, simply acting to bring the other agents together with a clear focus on physical redevelopment rather than community renewal.

The URCs have, however, been involved in very significant projects. *Liverpool Vision*, for example, one of the first URCs, has been involved in the dramatic transformation of the Ropewalks district into a cultural quarter as well as the vast Paradise Street redevelopment in the commercial core.

The ODPM (2004b) undertook a policy stocktake of the URCs and concluded that the programme should continue. There has, however, been some tinkering over the details of implementation, particularly over the working relationship between the URCs and their partner regional development agency and the extent to which English Partnerships would be involved. Although the programme has expanded from the 12 URCs originally envisioned in the Urban White Paper, it is interesting to look at the geographic locations of these bodies. With a few exceptions, the URCs are disproportionately concentrated in the northern former industrial heartlands, suggesting that these areas still face major challenges requiring state intervention.

Interestingly, however, in addition to the URCs, the Labour government has decided to revive the urban development corporation (UDC) model in certain circumstances. These are still operating under the original legislation, namely section 136 of the Local Government, Planning and Land Act 1980. This most Thatcherite of regeneration tools was set up with the intention of:

- bringing land and buildings into effective use;
- encouraging the development of existing and new industry and commerce;
- creating an attractive environment; and
- ensuring that housing and social facilities are available to encourage people to live and work in the area.

Notably absent from these aims is any sense that the UDC should attempt to foster community cohesiveness or actively work with existing communities within their area. This is particularly significant given the critique of the UDC model was always its divisiveness by effectively engaging in state-sponsored **gentrification**, building **infrastructure** to attract a new (wealthier) population, rather than engaging with needs of the existing community.

Community policies

Although this book focuses less on the community aspects of regeneration, it is interesting that there has been a real split in policy terms during the New Labour period. What has been produced is a division between community-led policies, which are broadly tagged with the label 'renewal', and changes to the physical infrastructure which is broadly referred to as regeneration. This changing discourse has broken up the notion of an holistic economy–society–environment conception of regeneration. In all fairness, however, one can perceive coordinating agencies such as the Housing Market Renewal Pathfinders (see below), as an attempt to draw together these different strands, although with varying degrees of success.

This idea that there is a somewhat separate discourse of community/social renewal comes out of the Neighbourhood Renewal Unit (NRU) which was set up in 2001. The associated Neighbourhood Renewal Fund (NRF) provided £1.875bn in 2001–06 to 88 of the most deprived authorities in England, with a further £1.05bn allocated in 2006–08 to the

86 most deprived authorities (Neighbourhood Renewal Unit, 2007). Perhaps the most significant initiative operated by the NRU is the New Deal for Communities (NDC), though this programme slightly predates the NRU having been established in 1998. The idea is that NDC partnerships are set up at local level to produce a local response to tackle five key indicators of social deprivation: unemployment, crime, educational under-achievement, poor health and problems with housing and the physical environment (although the NDCs do not have major resources for large-scale rebuilding programmes).

The NDCs have received quite a critical reception. For all that the intention is that targets and actions should be set locally, there was considerable underspend of resources as locally agreed targets were subsequently rejected at national level. As Imrie and Raco (2003: 27) argue: 'Communities are often "shoehorned" on to local policy initiatives according to central government guidelines … limiting the effectiveness of programmes on the ground.' There remains a tension in policy on community renewal between the rhetoric of bottom-up community empowerment and the setting of very rigid, centrally-driven priorities for what issues can and cannot be tackled.

This continuing tension in community policy is reinforced by the creation of Local Strategic Partnerships (LSPs). Where the NDCs are targeted to specific areas – a group of housing estates, for example – the LSPs take a larger-scale overview. Originally LSPs were limited to the areas which qualified for NRF resources, but this quickly expanded to include most areas of England with LSP boundaries matching those of local authorities. The core group of LSPs have considerable power as the main conduit through which Neighbourhood Renewal Funding passes into those 86 most deprived local authority areas. These bodies are not, however, democratically elected, but are instead run by representatives from partner organisations, particularly local authorities, local police authorities as well as the health and education sectors, alongside a variety of other state and non-state actors.

LSPs work around the notion of 'floor targets', a term established in the Treasury's Spending Review of 2000 to set minimum standards on a variety of social indicators for deprived areas in order to narrow the gap between these and less deprived areas. This type of indicator-driven target setting has become a familiar part of Labour policy-making over the last decade and the floor targets have become a key part of the LSPs' operation (Bailey, 2003). Indeed, more recently, the idea of floor targets has been still more formalised into Local Area Agreements (LAAs). From 2007 LSPs were required to operate through LAAs, which essentially represent an agreement between central government, the local authority and the LSP as to what the priorities for action are in a given area. While the rhetoric of joined-up thinking between different agencies is laudable and ensuring that socially deprived areas are targeted for improvements in education, health and public safety entirely sensible, the extent to which LSPs will truly respond to local needs, rather than chasing floor targets set nationally, is moot.

Commission for Architecture and the Built Environment

With a parallel 'renewal' agenda covering social policy, 'regeneration' can be seen as somewhat skewed towards a concern with physical infrastructure. This said, when Labour came to power in 1997, the country clearly had a distinct problem dealing with a very troubled legacy of post-war urban design which needed to be addressed. Just as the Urban Task

Force had a distinct design-led flavour, there were other reforms in the first New Labour term that emphasised the importance of good quality architecture and spatial planning in Britain's cities. The Royal Fine Art Commission had been set up in 1924 and had the power to call in and comment on development plans – though it had no statutory power to enforce changes. In 1999, the Commission was rolled into the newly established CABE (Commission for Architecture and the Built Environment). As with the Urban Task Force, CABE was closely associated with the personal interests of the Deputy Prime Minister John Prescott, and its creation brought with it a clearer remit to promote high-quality design, both through commenting on major development plans and providing advice to developers and various public bodies. As with the original Royal Fine Art Commission, however, CABE's advice is not statutorily binding and its main power is in naming and shaming poor design through its Design Review Comments. In recent years CABE has helped establish the principle of design coding, whereby designated redevelopment areas have a series of design specifications at various different degrees of detail laid out for them. These can include such things as height of buildings, set backs from the street, overall street/frontage patterns, guidance on material textures/colours, even sometimes specific guidance on detailing. The idea was piloted in a series of projects in 2004–06 and was subsequently embedded into *Planning Policy Statement 3: Housing* (CLG, 2007a). There will, therefore, be a much greater use of design codes in the future, with CABE positioned to provide advice and guidance on how they are drawn up.

Construction policy

Where the Urban Task Force chaired by Lord Rogers grabbed most of the headlines, it was Sir John Egan's Construction Task Force which has probably had more of a direct impact on current policy. Sir John was not a construction industry insider and his review highlighted concerns with the flexibility of the building industry, in particular how easy it was to introduce new practices and new technologies as well as the kinds of training needed by construction workers and managers to help meet these new challenges. Egan particularly identified the advantages of longer-term partnerships between construction firms and developers, noting the cost and quality advantages that these arrangements brought to the housing association sector (Construction Task Force, 1998). The original review has fed into a broader Egan agenda supported through a new executive agency, Constructing Excellence, co-funded by both the CLG and the DBERR. The construction industry has a massive role to play in meeting targets on sustainability because of the very large impact that construction has on the environment. By helping to restructure how the industry operates, the Egan agenda has encouraged the much wider application of new technologies and practices which has driven more sustainable construction.

This agenda of sustainable construction has also been driven by a progressive tightening up of the Building Regulations, with ever more stringent standards of insulation and energy use. The Regulations have also been altered to minimise other environmental impacts of new construction, such as reducing the quantity of surface water runoff through the use of sustainable drainage systems (SuDS). The Building Research Establishment has also been driving this agenda through its Environmental Assessment Method (BREEAM) and its EcoHomes standards. Many of these ideas have subsequently been absorbed into the CLG's Code for Sustainable Homes, launched in 2006. Level 6 of that Code is for homes which are effectively carbon neutral, in that they generate sufficient energy from renewable sources to 'pay back' any energy they draw from the national grid. English Partnerships is currently running a Carbon Challenge on behalf of the CLG to encourage the construction of carbon neutral homes, with ideas being mooted of all new homes having to be carbon neutral by 2016 (English Partnerships, 2007). This poses as massive challenge to the construction industry and gives some indication that the government is now taking the threat of climate change very seriously in terms of policy practice.

The Sustainable Communities Plan

The ODPM/CLG's most visible intervention in the physical environment of England is the Sustainable Communities Plan. The plan was launched in February 2003 and led with a strong physical reconstruction remit, outlined in its first publication, *Sustainable Communities: Building for the Future* (ODPM, 2003). The priority areas were identified as:

- the 20% most deprived wards in England;
- former coalmining areas;
- growth areas in the south east (Milton Keynes and the south Midlands, the London–Stansted–Cambridge–Peterborough Corridor, Thames Gateway and Ashford);
- the northern growth corridor;
- strategic areas of brownfield land; and
- the Housing Market Renewal Pathfinder areas.

This document was not uncontroversial, not least because it proposed an additional 200,000 homes in the 'growth areas' in the south east of England, most notably in the Thames Gateway area. To many critics, the Communities Plan therefore gave a mandate to concrete over the south east. A key agency in the delivery of the plan is English Partnerships, which was founded in 1993 as the 'national regeneration agency'. It has grown significantly, both in remit and power, absorbing a number of other agencies including the old Commission for New Towns, the defunct UDCs and the remaining assets of the Housing Action Trusts. It has ended up as a major landowner in several strategic areas and thus critical to the delivery of major changes in physical infrastructure.

A key delivery mechanism for the Sustainable Communities Plan are the Housing Market Renewal Pathfinders. Nine areas in the Midlands and north of England were designated in 2002, where it was deemed that the housing market was near collapse with abandoned houses and a decayed physical environment. The Pathfinders began operation in 2003 with a remit to restart the housing market by making the area more attractive through physical reconfiguration. On the one hand, this is quite an enlightened initiative, recognising the very different challenges faced in certain parts of the country compared to the high-demand south east (see Box 2.2). In the period 2004–08 £1.2bn has been allocated to the Market Renewal Fund, with Pathfinders also expected to draw in resources from the Neighbourhood Renewal Fund and the New Deal for Communities, both of which are targeted at the 'social' side of regeneration. Stuart Cameron (2006) has argued, however, that the Pathfinder initiative has shifted from a particular concern with housing abandonment to a more general drive to 'modernise' housing areas, i.e. fitting in with broader regional economic policy rather than prioritising the particular needs of individual communities/areas. CABE's (2005a) review of how the Pathfinders were progressing, was very much focused on the need to work at a sub-regional level, to consult carefully with communities to work out the source of problems in particular areas. This can be seen in part as a reaction to local criticism that the Pathfinders have been rather insensitive about the way that certain properties have been demolished. Any programme which seeks to physically reconfigure an area and bring in new residents is always open to the charge that it represents little more than gentrification and this accusation has dogged the Pathfinders.

Box 2.2   Bridging NewcastleGateshead Pathfinder

One of the nine Housing Market Renewal Pathfinders established by the Sustainable Communities Plan, Bridging NewcastleGateshead (BNG) covers the inner urban areas of both Newcastle and Gateshead. The conurbation is still feeling the effects of post-industrial decline with economic growth at around 1.3% compared to a national average of 3.1%. The BNG covers an area containing 140,000 people and 77,000 dwellings, of which 47% is socially rented and 40% owner occupied. Vacancy levels run at 7%, which is relatively high and the area experienced a 6% population decline between the 1991 and 2001 censuses (Leather et al., 2007: 134).

During the first three years BNG oversaw the refurbishment/upgrading of 800 homes, particularly targeting the 'Tyneside flat', a housing type peculiar to the north east of England, converting these into single family homes. A further 50 empty properties were brought back into use, while some 1,300 'obsolete' properties were demolished to make room for new properties to be built. The BNG also boasts that during its first three years private developers have delivered 1,850 new properties and that land for a further 600 new dwellings has been acquired (Coulter et al., 2006: 23).

The Sustainable Communities Plan has had a general concern with capacity-building among the different organisations drawn into the regeneration process. Following on from recommendations in the Urban Task Force Report, the Urban White Paper (DETR, 2000) asked the RDAs to establish 'Regional Centres of Excellence' which would seek to foster a skills and training culture for issues around the built environment. The idea was raised again in *Sustainable Communities: Building for the Future* (ODPM, 2003) and the RDA Advantage West Midlands was the first to establish such a body – RegenWM – in 2003. The Egan Report on *Skills for Sustainable Communities* (ODPM, 2004a) gave further impetus to this idea and there are now eight such centres of excellence. The idea is not that these bodies should supply training, but that they should facilitate groups in the private, public and community sectors, identifying the skills they need to work in regeneration and helping them to acquire the necessary training. The centres of excellence are supported by the Academy for Sustainable Communities (ASC). This national body is charged with increasing the skills base in the sector in order to improve the kinds of new environment that are produced through regeneration. As with the regional centres of excellence, the ASC is not intended to provide training itself, but rather to foster a culture within the sector which appreciates the importance of skills. The ASC also took over some of the awards schemes which were run by the ODPM, acknowledging innovation and best practice, launching the Future Vision awards in 2007 for undergraduates coming up with ways to improve community sustainability.

Key points

- Central government reorganisations have somewhat fractured the notion of holistic, sustainable regeneration, not least through the economic development remit being passed to the Department of Business, Enterprise and Regulatory Reform, while social/community issues have been shunted into a separate discourse of 'renewal'.
- The Treasury-driven Barker reviews have placed a great deal more emphasis on market forces determining how land should be developed and this has subsequently fed through into planning policy.
- Regional policy in England has significantly beefed up planning powers and powerful, well-funded development agencies but has been compromised by a lack of democratic accountability.
- The Sustainable Communities Plan is driving large-scale building programmes, particularly in the south east, but also through demolition and reconstruction in the Pathfinder areas. This has been perhaps the most controversial element of the ODPM/CLG's work.

Devolution

Perhaps the most significant long-term shift signalled by the 1997 election was devolving a degree of power to Scotland, Wales and Northern Ireland, something which had been fiercely resisted by Conservative administrations. Unlike the subsequent botched attempt at creating regional government in England, devolution to the Celtic fringe proceeded very smoothly, in part perhaps due to the fact that real power was on offer. Each of the three regions had previously been controlled by a dedicated government department at Westminster and although there was some variation from English policy – particularly in Scotland – there was an overall coherence. Since the new Welsh and Northern Ireland Assemblies and Scottish Parliament have been established, there has been more of an opportunity for regionally distinctive policies for urban regeneration to be devised and this has happened to a greater or lesser degree. This is not to say, however, that there is no cross-fertilisation of good practice, for instance with Scotland having recently chosen to adopt the urban regeneration company model from England and Wales.

Scotland

Of the three new governments set up after 1997, the Scottish Parliament overseeing the Scottish Executive is by far the most powerful. In the referendum, Scots were not only given the opportunity to vote in favour of devolution itself, but also whether a devolved parliament would have power to vary the rate of taxation within Scotland. Scots voted in favour of both and since 1999 Scotland has had a measure of independence though not sovereignty. There was always the potential for urban policy to be disproportionately influenced by the demands of the central belt conurbation and the country's two major cities, Glasgow and Edinburgh. Together, the central belt comprises half of the country's population and economic activity (Bailey and Turok, 2001) as well as providing the power base for the Labour party. From 1999 to 2007, however, Scotland was governed by a coalition of Labour (solid in urban areas) and the Liberal Democrats (strong in rural areas). This coalition has meant that there has needed to be a more pragmatic, consensus approach to all aspects of policy and much less dominance by the central belt. It also, from the beginning, marked a distinct break from the Labour-dominated government in Westminster. The narrow victory of the Scottish National Party in 2007 means that the Parliament continues to be run by a minority administration, with the need for a consensus approach to government continuing.

Where the Westminster-controlled Scottish Office had previously set up arm's length bodies to tackle aspects of regeneration, the new Scottish Executive has brought some of these functions back under direct control. Scottish Homes, for example, was established in 1988 as a non-departmental public body, with significant central government funding to build social housing and foster social regeneration schemes. In 2001 the main functions of Scottish Homes were absorbed into a new body, Communities Scotland, which is directly overseen by the Scottish Executive Development Department. Where English Partnerships is merely the 'national regeneration agency', Communities Scotland adds the word 'housing' to its remit and has six objectives:

- To increase the supply of affordable housing where it is needed most.
- To improve the quality of existing houses and ensure a high quality of new build.
- To improve the quality of housing and homelessness services.
- To improve the opportunities for people living in disadvantaged communities.
- To support the social economy to deliver key services and create job opportunities.
- To use our experience of delivering housing and regeneration programmes to inform and support the development of Ministerial policies. (Scottish Executive, 2007).

Affordable housing is a major issue in Scotland, given the high proportion of the population living in social housing, especially within Glasgow. The very large post-war estates that were built by councils have become notorious for chronic, multiple deprivation compounded by crumbling housing infrastructure. Communities Scotland thus ties together issues of providing high-quality, affordable housing with tackling the major social problems in run-down parts of Scottish cities – arguably demonstrating rather more joined-up thinking than English Partnerships.

Communities Scotland was responsible for the operation of the Social Inclusion Partnerships (SIPs). Of these 48 bodies, 34 were specifically established to cover a geographic region with the intention that they should coordinate with public and other bodies in that area and 'bend' their resources towards a social inclusion agenda – thus having some parallels with the Local Strategic Partnerships in England. In many cases, the local authority was a key partner in the SIP process, but there was also an emphasis on capacity-building for much greater direct community involvement (Lloyd, 2002). From 2003 the SIPs began to be integrated into the Community Planning Partnerships (CPPs) and in 2004 the funds for SIPs were amalgamated into the Scottish Executive's Community Regeneration Fund. This is a three-year project (2005–08) designed to help the most deprived 15% of neighbourhoods. The CPPs were asked to draw up Regeneration Outcome Agreements, making a commitment to improve indicators of social exclusion in their areas, particularly on health, education and employment. Again, while there are clear parallels to the English LSPs, this is a particularly Scottish response – although the Scottish Executive, it seems, is not immune from the habit of the Westminster government of rebranding its funding schemes every few years and creating a whole new set of acronyms to learn.

Devolved government does not mean that Scotland ignores trends in contemporary policy circles, just that the details of application can differ. A good example of this is with the notion of cultural clustering. The economic arguments behind this will be discussed in more detail in Chapter 4; put briefly, there is a belief, influenced by the ideas of Richard Florida (2002), that cities in which both traditional and alternative cultural resources abound will attract economic growth through the 'creative' industries. Cities and regions across north America and Europe have thus been looking at ways of nurturing these cultural resources. The Scottish Executive has produced a National Cultural Strategy which seeks to encourage the growth of cultural clusters within Scottish cities, both facilitating existing ones, such as in Glasgow, and developing new ones, such as in Dundee. While this policy is not without its critics (see, for instance, McCarthy, 2006), it does reflect an attempt to give a regional spin to a 'fashionable' policy idea – something that has been facilitated by devolution.

There have also been circumstances where the Scottish Executive directly takes on a policy structure from England. In terms of urban regeneration the most noteworthy adoption has been that of the urban regeneration company (URC) model. Five URCs have been established in Scotland following an evaluation of the English and Welsh URCs, which indicated that the private sector was more willing to invest in these areas because the presence of a URC demonstrated a willingness by local public sector actors to work together in a coordinated way. Riverside Inverclyde URC, for example, is currently promoting a significant transformation of the old industrial waterfront and, as a result, Inverclyde was named fourth in the top ten property 'hot spots' in Scotland for first-time buyers in a survey by the Royal Bank of Scotland (2006).

Wales

Where Scots were given the opportunity to vote in favour of setting up their own Parliament, Welsh voters were only offered an Assembly, with slightly fewer powers. Furthermore, the Welsh were not given the opportunity to vote on whether the Assembly would have the power to vary taxation within the Principality. Indeed, where the Scots were given powers over primary legislation in most areas of domestic policy, the Welsh Assembly merely makes subordinate legislation which allows variations to Acts of Parliament in Westminster.

A result of this is that urban regeneration policy in Wales has generally not deviated from the English model to the extent that has happened in Scotland. Prior to 2007 the government of Wales was somewhat unusual in that the executive and legislative functions were contained within a single body, which have since been separated out into the Welsh Assembly Government and the Welsh Assembly respectively. The Welsh Development Agency was absorbed into the executive branch in 2007, becoming part of the Department for Economy and Transport. This reform is a significant departure from the English and Scottish model of having a regional development agency with an economic remit separate from the elected government.

Interestingly, therefore, this economic development aspect was not covered by the Department for Social Justice and Local Government, which has the specific regeneration and housing remit (although the Department for Economy and Transport also has a regeneration team). Indeed, planning functions are separated out into the Department for Sustainability and Rural Development. In the Welsh Assembly context, 'regeneration' seems to refer primarily to measures to promote social inclusion/community cohesion. This is not to say that major projects are not occurring which can be thought of as urban regeneration in the more general sense. Wales uses the same urban regeneration company (URC) mechanism as England and Newport Unlimited URC is currently coordinating a series of major projects to transform the central area of the city by 2020. The model of multiple partners coming together to deliver this quite radical reconfiguration of Newport is not that different from England except that a regional government department is one of the key funders rather than a regional development agency.

Northern Ireland

Northern Ireland is a slightly more complicated case, where before an Assembly could be set up, an agreement between the different political factions of this troubled province was necessary. The Belfast, or 'Good Friday', agreement paved the way to a referendum on a system predicated on the sharing of power between the largest parties of the Unionist and Nationalist communities. Urban regeneration is thus caught up in what is a unique political context in the UK, with very rigorous procedures to ensure that neither community is seen to be gaining an unfair advantage in terms of public policy and public spending. Assembly business was suspended between 2002–07, with protracted arguments between the different political parties over the direction of the peace process.

The prolonged period of suspension meant that regeneration in Northern Ireland perhaps lacked the drive and direction that politicians provide to the process elsewhere. The Northern Ireland Department for Social Development has responsibility for overseeing regeneration efforts. Interestingly, and perhaps as a consequence of the sensitivity surrounding the different communities within Northern Ireland, there seems to be a much greater integration of infrastructure and social projects. Even the Regional Development Office, whose mainland equivalents focus on an economic and physical remit, has interests ranging from comprehensive town centre redevelopment schemes to neighbourhood renewal projects.

In order to ensure that neither Protestant nor Catholic communities are seen to be benefiting 'unfairly' from initiatives designed to tackle social need, multi-criteria scoring has been introduced to set priorities. These scores are derived from census data and weighted according to a complex set of multi-scaled calculations which go right down to the Enumeration District level. Brian Robson of Manchester University produced the first of these scoring tools in 1994 and this was subsequently revised by Michael Noble of Oxford University in 2001 (Northern Ireland Assembly, 2002). The 'Noble Index' is thus a key tool in implementing the 2003 People and Place strategy for Neighbourhood Renewal. Thirty-six areas have been targeted for action based on their scores, mostly concentrated in the two major urban centres, with 15 in Belfast and six in Londonderry/Derry. People and Place stresses the importance of engaging local people with meaningful participation in the renewal process as well as attempting to ensure that the process does not exacerbate community segregation.

Evidence of this need to ensure meaningful community engagement was even apparent in the Laganside Urban Development Corporation, in spite of the UDC model being overwhelmingly driven by a physical infrastructure and economic development agenda. Laganside Corporation proudly boasted that between its inception in 1989 and ceasing operations in 2007 it secured over £900m of investment in the area.

At the same time, however, the Corporation also funded community projects within its area and its annual reports are run through with assessments of how the organisation and its actions comply with legal requirements to promote tolerance and equality. In Northern Ireland, community, politics and physical renewal truly cannot be considered separately.

Key points

- Devolution has resulted in regional regeneration powers being subjected to much closer democratic scrutiny.
- Although there are significant areas of common ground between the devolved governments and England – particularly over notions of sustainability – devolution has resulted in more regionally-specific policy directions.
- In all three regions there does appear to be a much closer integration of social policies with physical regeneration than is the case in England with its parallel discourse of community 'renewal'.

Further reading

A key problem with a chapter of this kind is that it dates rapidly. Government websites are one of the only ways to keep abreast of the latest policy initiatives, as well as providing relatively easy access to policy reviews commissioned from external agencies. For an insight into some of the ideas on good design, mixed use and sustainable communities which underpin the policy rhetoric, it is worth revisiting the original Urban Task Force report (1999), steeped though it is in the optimism of the early period of Labour government. Imrie and Raco (2003) give a good general overview of urban policy as well as giving good detail on the social/community aspects, while Cochrane gives a longer history of urban policy in the UK. Greenhalgh and Shaw (2003) give an interesting insight into the tangled world of English regional policy. Bradbury and Mawson's (2006) edited collection is a heavyweight review of the impact of devolution on UK policy, with particular sections on planning.

Bradbury, J. and Mawson, J. (eds) (2006) *Devolution, Regionalism and Regional Development: The UK Experience* (Routledge, London).

Cochrane, A. (2006) *Understanding Urban Policy: A Critical Introduction.* (Blackwell, Oxford).

Greenhalgh, P. and Shaw, K. (2003) 'Regional development agencies and physical regeneration in England: can RDAs deliver the urban renaissance?', *Planning Practice and Research*, 18(2–3): 161–178.

Imrie, R. and Raco, M. (2003) *Urban Renaissance? New Labour, Community and Urban Policy* (Policy Press, Bristol).

Urban Task Force (1999) *Towards an Urban Renaissance.* Final Report of the Urban Task Force chaired by Lord Rogers of Riverside (DETR, London).

3 Governance

Overview

This chapter explores ideas of governance which are used to understand the political processes through which urban regeneration is delivered. The following areas will be covered:

- *Definition: from government to governance*: examines what governance means and why it is important for urban regeneration.
- *Understanding the state*: breaks down the idea of 'the state' as a single body and looks at the complex arrangements of state institutions involved in cities.
- *Different forms of governance*: looks at the debates between scholars on how different relationships between the state and non-state sectors can be theorised.
- *The new institutionalism*: examines how the structure of institutions themselves can have a significant effect on how governance takes place.
- *Community involvement*: examines the place of the community in governance arrangements between state and non-state sectors.
- *Regional regeneration and the European Union*: highlights the growing importance of the EU for regeneration and explores whether this represents an undermining of national state power over the process.

Definition: from government to governance

In English, the word 'government' has a number of meanings, both as noun and verb. Government can be taken to represent the institution charged with the act of governing – where it becomes synonymous with '*the state*'. It can also be used to mean the actions of that institution – what the state seeks to achieve. Governance, in turn, refers to the process of *delivering* the aims of the state (Newman, 2001). The act of a government agency paying welfare benefits to the unemployed, for example, can be considered as governance, as it is delivering part of the state's social agenda.

Not all of the state's aims are delivered by its own agencies, however. Increasingly non-state actors are being brought in to help deliver services which the state does not wish to be directly involved with. One of the more controversial examples of this is the public–private partnership, where a private contractor bids for the right to build a new capital resource (for instance, a hospital) and lease it back to the public sector over a fixed period, while retaining responsibility for its maintenance. The public sector is then free to concentrate on more core activities with no need to have separate competence in construction and maintenance, which are already available in the private sector.

The concept of governance is therefore a very useful way of looking at a whole range of actors who are involved in the delivery of policy. This is particularly true for urban regeneration, where the idea that it should be directly delivered by the state alone is unrealistic because of its large scale and diverse remit as well as high costs. While the state sets a framework in which urban regeneration takes place, a large number of actors, from developers and construction firms through to charities and local communities, are involved in the actual process of undertaking a regeneration programme.

Understanding the state

A monolithic state?

As the discussion above suggests, the notion of governance raises all kinds of issues about 'the government' as an institution in itself. While an overarching concept like 'the state' includes all the formal institutions of government, it does tend to make us think of the state as a single entity. Clearly this is not the case. The state can be broken down into a number of different scales (local, regional, national) and into different types of institution, for example central government departments as against executive agencies which operate at arm's length from central government. Supranational government, in particular the European Union, are also of critical importance and further muddy the waters of the concept that the state is a singular, monolithic entity speaking with one voice.

Given that the direct effects of urban regeneration are felt most acutely at the local level, local authorities play a critical role in initiating and managing the processes. Even a local council is no coherent entity speaking with one voice. Regeneration projects need to coordinate between the different agendas of different council departments. The aims of an economic development department trying to maximise economic activity in a city may come into conflict with a planning department which has to enforce a set of planning rules laid down by central government.

Local councils are run by directly elected councillors who have a political mandate and direct accountability to their constituents. There can, however, be some tensions where a local council is controlled by one political party while central government is controlled by one of the opposing parties. There is an intermediate tier or regional government. The Scottish Parliament and Welsh Assembly are directly elected bodies while the regional assemblies in England are comprised of local authority representatives (at least 70%) and representatives from local 'stakeholders', such as local business interests, further and higher education, trades unions, and so forth. These differently mandated regional governments play a significant role in urban regeneration, in particular through acting as the regional planning bodies.

The original intention was that the English regional assemblies would gradually become directly elected bodies through a series of local referenda, though this came to a halt following the overwhelming rejection of the proposed body for the north east in November 2004. The lack of direct electoral mandate and the inclusion of 'stakeholder' groups makes these bodies rather interesting, particularly as the regional spatial strategies they oversee carry statutory weight. The assemblies also have responsibility for overseeing the regional development agencies – state agencies which attempt to foster economic growth in the regions – distancing these powerful bodies still further from democratic accountability.

Even before one considers national government, therefore, it is clear that 'the state' as it affects urban regeneration is an exceedingly complicated and tangled series of overlapping institutions with more or less accountability at the ballot box. Since the devolution of powers to Scotland and Wales in 1998 the British government based in Westminster has had much less direct control over issues relating to urban regeneration in these areas. The Department for Communities and Local Government (CLG) is the crucial department of state as far as England is concerned, laying down overall policy frameworks within which the other state institutions have to operate – for example, publishing the series of Planning Policy Statements, which govern the direction of planning policy. CLG also provides the funding to the regional development agencies and has several executive agencies, such as English Partnerships (the national regeneration agency) and the Commission for Architecture and the Built Environment (CABE), which looks at improving urban design quality.

Hollowing out?

Bob Jessop (1994), examining the changes to the political system in the post-war period, talks about the state being 'hollowed out'. National governments still carry the appearance of having their powers intact but, in fact, aspects of their responsibilities have been removed (like hollowing a log) through being passed upwards to supranational bodies (for instance, the European Union) and downwards to local and regional governments. This process is sometimes referred to as *glocalisation* (Swyngedouw, 2004) and has significant implications for the governance of regeneration in the UK. Britain has traditionally had a very strong central government, something which was reinforced during the 1980s under Margaret Thatcher's government in an attempt to undermine the power of local authorities. Continental Europe has a much stronger tradition of regional government, and funding streams established by the European Union tend to bypass national governments. European resources became an increasingly important component of funding for local regeneration projects through the 1990s (see below).

European Union funding is tied to certain criteria, which often involve levering in private sector resource and meeting certain targets in terms of involving a range of local stakeholders. This means that the relationship between different parts of the state and different non-state stakeholders has become critical in urban regeneration. The idea of governance is thus exceedingly useful as it provides a framework for understanding the relationships between the different actors involved in the delivery of policy aims.

The shadow state

Despite the fact that the state is a multifaceted institution and does not speak with one voice, there are concerns that where the state enters into relationships with non-state actors, the aims of the state will dominate. This is particularly the case where the state hands responsibility for delivery of services over to charitable and voluntary groups – sometimes referred to as the 'third sector'. Such groups can become particularly dependent on that state funding stream and thus vulnerable to having their core interests shifted in line with what the state wants – the threat being that funds will be withdrawn otherwise.

Jennifer Wolch (1990) used the term 'shadow state' to describe this phenomenon, where actors in the 'third sector' (i.e. neither public nor private) are captured by the state while formally remaining separate from it. In urban regeneration, a good example of this is the housing association or registered social landlord (RSL) sector, which provide affordable housing to people on lower incomes. There is a large amount of funding available for RSLs from the Housing Corporation – an executive agency of central government. This funding is dependent on those RSLs delivering certain policy aims, such as the kinds of houses they build, where they build them and the environmental standards with which they comply. Indeed, there has been some argument that the need to comply with performance indicators, laid down by the Housing Corporation as part of a broader trend towards target-driven management in the public sector, has in fact been harming the business performance of RSLs and their social mission (Sprigings, 2002).

In terms of the governance of urban regeneration, therefore, there are two main fears for the charitable/community sector. The first is that while they might have been brought on board to meet funding requirements for inclusion, their voices will simply be drowned out when big decisions over resource allocation are being made. The second, and perhaps more pernicious, is that those organisations will be 'captured' by the way they have become involved with the process and the source of their funding. Reproducing the aims of the state agents, becoming their shadow, these organisations are moved away from their core mission and, indeed, can stop playing the role they were nominally brought in to play – i.e. to provide a different kind of input to the project.

Key points

- The state no longer seeks to deliver all its policy aims directly through its own agencies.
- 'The state' is no single entity, but speaks with numerous voices with different interests.
- The development of European-level government and increasingly powerful regional government has led to arguments about the power of the central state being 'hollowed out'.
- A criticism of using charitable 'third sector' bodies to deliver policy is that those bodies can end up merely reproducing or 'shadowing' the policies of the state, losing their individual expertise.

Different forms of governance

A great deal of work on what governance meant at a local level was undertaken during the 1990s in a programme of research funded by the Economic and Social Research Council (ESRC). In critiquing some of this work, the public policy expert Jonathan Davies (2001) came up with a useful typology of different forms of governance: governance by government, governance by partnership and governance by regime. To these we can add governance by networks, an idea associated with that ESRC research programme and particularly R.A.W. Rhodes (1997). Each of these is explored in turn though it should not be assumed that one is necessarily 'better' than another. Different forms of governance may be more or less appropriate depending on what the state is attempting to achieve in a given set of circumstances.

Governance by government

Governance by government is perhaps the most straightforward model. Essentially, the aims of government are delivered by the government itself. As discussed above, governments are very large institutions operating in a variety of guises and at different geographic scales, hence it can be appropriate for different parts of the state, with different remits, to work together on particular projects.

In the redevelopment of cities, this model tended to apply during the period of the post-war reconstruction up until the early 1980s. At this time local councils had considerably more self-determination when it came to decision-making and spending than they do today. Local councils are considered as part of 'the state', as it is broadly defined, and the working relationship between central and local government fits the governance by government model. Major housing and reconstruction programmes undertaken in the 1950s and 1960s were carried out by local authorities through grants and subsidised loans provided by central government. Indeed, many local authorities even directly employed their own teams of construction workers to carry out some of the building work.

The notion that the state should do this kind of work itself is clearly affected by the socialist ideals which were a strong undercurrent in the post-war reconstruction. There were, however, undeniable problems with inefficiency and corruption in some, though by no means all, parts of the post-war programme. High-profile scandals included T. Dan Smith, former Chair of the Housing Committee in Newcastle, and Alan Maudsley, former Birmingham City Architect, who were both jailed on separate corruption charges in the 1970s. These scandals gave ammunition to those who opposed the socialist model of state provision. The Conservative central government under Prime Minister Margaret Thatcher, which was elected in 1979, broke the post-war political consensus. Thatcher argued that the state was an inefficient means of delivering services – a belief which was most famously acted upon in the privatisation of many state-run businesses, including the major utilities.

The Thatcherite attack on the state as service provider went beyond the waves of privatisations. Local authorities, condemned as corrupt and inefficient, had severe restrictions imposed on them, determining how they could raise and spend revenue. The big post-war state housing programmes were brought to an end through the slashing of grants and subsidies to local government. Clearly, the state did not completely stop delivering services under Thatcher, but there was a clear ideological shift away from the idea that the state should deliver its own aims wherever possible to one in which it was felt that the state should only be involved in those areas where other sectors of society could not deliver that service better. As a result, the governance of the UK was broadened out from the governance by government model.

Governance by partnership

The second model of governance discussed here is where the state brings in partners to take some of the responsibility for delivery. If, under Thatcherism, the government was not going to deliver on all its policy aims itself, then external agents had to be brought in. This change of emphasis should be taken alongside a growing notion that urban *redevelopment* needed to be thought of as a more all-encompassing, **holistic** model of *regeneration*, acknowledging that the city is a hugely complex entity with a diverse range of problems which the state acting alone could not remedy. Non-state actors, in both the private and charitable sectors, were acknowledged as having a variety of expertise and resources which could be productively harnessed. In the 1980s, most of the emphasis was on drawing in private sector finance, with an increased recognition of the usefulness of the charitable sector emerging during the 1990s.

Though the idea of securing private sector investment came to prominence under Thatcher, that is not to say that it was entirely absent before. When redeveloping Britain's city centres, local councils had long been working in partnership with development companies and other private actors to deliver new shopping centres and other resources. In the 1980s a language of *levering in* private finance emerged, using public money to attract investors to areas which would otherwise pose too high a risk to developers. This was the era of the urban development corporation (UDC), well-funded public sector bodies operated at arm's length from the central government which used a variety of economic incentives, particularly major **infrastructure** programmes combined with tax breaks to bring developers into run-down areas. The London Docklands Development Corporation, which operated from 1981 to 1998, invested heavily in infrastructure and was involved in building the Docklands Light Railway, 115 kilometres of new roads and reclaiming 826 hectares of land and water in the area. The construction of the now hugely economically successful Canary Wharf development, which was seen as somewhat high risk at the time, created a major new centre of business and professional services in an area of London which had suffered major job losses during the recession of the early 1980s.

The **Thatcherite** model of partnership was very much driven by the belief that the private sector, through the regulating mechanism of the market, knew best and that if a scheme was good for business, then it would be delivering on the aims of the government. While schemes like Docklands did a lot to regenerate physical infrastructure, they can be criticised for not having tackled some of the more intractable social problems in Britain's inner cities. Merely bringing in private sector finance, therefore, was not enough to achieve a more all-encompassing *regeneration* of an area – former dock workers were unlikely to find employment in the finance houses attracted to Canary Wharf. During the 1990s there was an increasing recognition that alternative actors needed to be brought into the process. Charities, voluntary bodies and the local communities themselves began to be seen as partners in the regeneration process.

The increased inclusion of the charitable sector in the 1990s does not imply that such organisations began to dominate the process – governance 'partnerships' are not an exchange of equals. Davies (2001), in fact, uses the word 'partnership' to imply that the state remains the dominant actor in this mode of governance. The 1990s were significant for a swing back towards involving local authorities – still very much part of the state – in regeneration. One of the reasons the UDCs had been set up was because the Thatcher government did not trust local authorities to deliver its policy aims. Indeed, there was a hope in the 1980s that regionally based Training and Enterprise Councils (TECs), would take the lead on local regeneration schemes, effectively bypassing local authorities. The TECs were designed to build relationships with local business and voluntary agencies to tackle unemployment and skills shortages. In practice they were not equipped to lead regeneration schemes. The Major and Blair governments of the 1990s took a much less hardline approach to local authorities, accepting that they had considerable expertise and local knowledge which made them well placed to coordinate locally-based regeneration projects.

Governance by networks

It was in this context that the local governance programme of the Economic and Social Research Council was operating during the mid to late 1990s. The leading figure in this programme was R.A.W. Rhodes, who defined governance as '*self-organizing, interorganisational networks*' (Rhodes, 1997: 53, original emphasis). The word 'network' is quite useful as it suggests the bringing together of a series of different actors to accomplish a task, though not actually putting together a single overarching body – each of the different actors remains independent. In this context, the 'network', as distinct from 'partnership', implies that the state does not dominate the process. In practice, however, weaker actors, such as community groups, tend to have much less influence over the process than those actors bringing political power or financial muscle to the project.

The distinction between networks and partnerships is somewhat blurred – just how much state involvement in a project transforms it from a network into a partnership? It is perhaps not worth getting caught up in the semantic argument. Instead, try to examine just what powers the state has over a particular regeneration project – it clearly remains a very important player, although it is important to re-emphasise that 'the state' is no single entity. Certainly, central government still provides a great deal of funding for regeneration projects, both directly and through its executive agencies such as the regional development agencies and English Partnerships, the national regeneration agency. Similarly, the involvement of local authorities in individual schemes gives the state considerable control over the process. Indeed, one can critique Rhodes' notion that these networks are 'self-organising' by the fact that it is often the state, at either the local level or through an executive agency, which brings together a network to produce a particular project, coalescing around a funding stream related to particular state aims. This was certainly true of the Single Regeneration Budget (1994–2001), where projects were generally led by local authorities, but when bidding for these resources they were required to demonstrate that a network/partnership had been set up with a range of actors to create a more inclusive project (Box 3.1). As a result, in terms of the actual implementation, the introduction of a network of non-state actors into the decision-making and delivery process means that unexpected outcomes are inevitable.

> **Box 3.1   Extracts from the guidance given to organisations bidding for the Single Regeneration Budget round 6 (CLG, 2006c)**
>
> The SRB is a flexible programme which supports schemes which can have a mix of the following objectives:
>
> - improving the employment prospects, education and skills of local people;
> - addressing social exclusion and improving opportunities for the disadvantaged;
> - promoting sustainable regeneration, improving and protecting the environment and infrastructure, including housing;
> - supporting and promoting growth in local economies and businesses;
> - reducing crime and drug abuse and improving community safety. (CLG, 2006c)
>
> Bids must be supported by partnerships representing all those with a key interest from the public and private sectors and from local voluntary and community organisations. The make-up of partnerships should reflect the content of the bid and characteristics of the area or groups at which it is aimed. In some cases, existing partnerships will put forward bids; in others, new partnerships (or adaptations of existing ones) will be formed. (CLG, 2006c)

Governance by regime

The idea of urban regimes was developed in the United States through the work of Stephen Elkin (1987), examining the city of Dallas, and Clarence Stone (1989) looking at Atlanta. Both authors examined in great detail the development of a long-term relationship between the city council and local businesses to promote the economic development of the city as a whole. A lot of this activity coalesced around landownership and development, but also took in social issues. For example, business leaders in Atlanta were active in pressing the city authorities to desegregate its schools in 1961, feeling that this would be good for business in the longer term.

Urban regimes, in Stone's model, work at a macro scale of policy – such as race relations – rather than on specific detail. By definition, this is a long-term type of relationship between the local state and business interests. There have been various attempts to apply the idea of regimes to the UK. There are, however, some significant differences between the political model in the UK and USA, particularly in the relationships between local councils and local business leaders, which tend to be less overt in the UK. The involvement of non-state actors in urban regeneration in the UK tends to take the form of medium-term relationships based around specific projects. In terms of regeneration, there does not seem to be the same kind of long-term strategic overview involving the private sector which could be characterised as forming a regime-type relationship. Looking for examples of regime governance in urban regeneration would seem, therefore, to be a bit of a red herring.

Case study: Park Central, Birmingham

A former council estate, butting up against the southern edge of Birmingham's central business district, Park Central is a good example of the kinds of innovative governance arrangements which are being demanded by the contemporary British state. The initial impetus for the scheme, unusually, came from a highly coordinated campaign by local residents who wanted to see some money being spent on their estate to improve its run-down environment. The Labour-controlled local authority initially resisted a proposal to apply for a central government funding stream which would have moved the estate into the ownership of a third-sector housing association. The **New Labour** government elected to Westminster in 1997 did not scrap the scheme and the local authority, somewhat reluctantly, agreed to authorise a bid after all. This scheme, the Estates Renewal Challenge Fund (ERCF), gave the housing association ownership of the properties and their curtilages as well as a large grant to meet refurbishment costs. The council's bid to the scheme was approved by the government and tenants voted in favour of leaving the public sector.

When the estate was in council ownership it clearly represented the governance by government model. The fact that local authorities have been starved of funding for maintaining their housing stock – and prevented by central government from independently raising revenue to do this – meant that this governance arrangement has become increasingly problematic. Schemes like the ERCF *forced* local authorities into a different mode of governance for housing, accepting that central government wanted social housing services in their cities to be moved into the partnership mode.

The housing association that was set up to take control of the housing stock in the area was called Optima Community Association. This stress on 'community' was deliberate, emphasising that the organisation was more than simply a housing provider, but was trying to improve all aspects of community life. The separation from the local authority was not a clean break, however. Several of the senior staff positions were recruited from inside the city council's housing department. At the same time, while Optima took ownership of the properties and their curtilages, the city council retained ownership of the public realm – not only roads and pavements but also a very large park which ran through the area.

This land ownership question became significant as Optima looked at ways to raise revenue to build new homes (the ERCF only paid for refurbishments). Optima were always going to demolish some properties and thus had land available to sell to a developer. In an outcome which was never anticipated when the original bid for ERCF resource was made, Optima and the city council cooperated in producing a more integrated masterplan for the area. They changed the size and shape of the park, demolished more property than originally anticipated and created a sizeable development parcel. Rather than selling off land piecemeal, the idea was to get a development company to come on board as a partner in redeveloping the site as a whole.

On some levels Optima had attempted to distance itself from the city council when attempting to bring developers on board. An indicative masterplan was commissioned by Optima to give developers some idea of the kind of scheme they wanted, rather than simply giving out copies of the city council-branded supplementary planning guidance. The development was thus distanced from the bureaucratic control of the local authority and the poor reputation of council estates. As one senior manager for Optima put it: 'Having worked on the outside, I didn't particularly want the whole area branded Birmingham City Council from a sales point of view – a development sales point of view. Because having worked on the other side, it's kind of off-putting' (interview, 9 August 2004). At the same time, however, it was made clear to developers that with the city council having a landholding interest, the partnership would be able to access the city's powers of compulsory purchase to smooth the process of land assembly.

While the Park Central scheme was a response to general aims of both the local and national state to improve the area, when it came to the detail Optima produced something quite different from what was originally envisaged by the remit of the ERCF programme. Optima acted as the centre of a network of agents acting somewhat independently of the state. At the same time, however, Optima ensured that this network was sufficiently closely tied in to state mechanisms, particularly at the local government level, to ensure that it could access some of the state's powers – particularly over compulsory purchase and land assembly.

After a lengthy bidding process, Crest Nicholson were chosen as the development partner and they quickly drew up a masterplan for the site. As the scheme has progressed, Optima has received income from land sales to the developer. In addition, the newly built park has included facilities requested by the community and Optima has received a proportion of new residential units to let out to new tenants (Figure 3.1). While by no means perfect, the scheme has had significant successes, which can in part be put down to the cooperative and mutually beneficial relationship between the different partners.

The relationship between Optima, its tenants, the city council and the developer is a medium-term one, coalescing around a particular project – Park Central. As such it does not possess the characteristics of a regime; it is neither sufficiently long term, nor is it looking to the more general interests of the wider city's development. It does, however, provide an example of how a partnership arrangement can work, both through making use of the state's power, but at the same time going beyond the specific remit of funding schemes and accessing the ideas and expertise (and resources!) of external agents. (For more details of this case study, see Jones and Evans, 2006.)

Key points

- A series of different models have been devised to describe the kinds of relationship between the state, the non-state sector and the delivery of policy, from governance by government (where the state delivers its own services) to governance by regime (where the state and non-state sectors work together in a very long-term, stable relationship to promote mutual interests).
- The debate over the precise role of the state in governance arrangements is very contentious, with Davies (2001) in particular arguing that in the UK the state remains a dominant *partner* in the regeneration process rather than behaving as merely one actor in a network of involved parties.
- In the UK today, regeneration schemes can only take place where a variety of different actors from both the state and non-state sectors are involved, although in practice it is often state actors which take the lead.
- The Park Central case is relatively unusual in that it is led by a third-sector body. Although it originated from a city council project and was originally funded by a national state scheme, it has established considerable independence from the regeneration programme which was originally envisaged. This demonstrates the problem of attempting to apply just one model of governance – the real world has a tendency to be messy!

Figure 3.1 The transformation of an inner city estate. Top, Lee Bank with low quality park and high-rise, 2000. Bottom, now rebranded 'Park Central', new flats surrounding a high quality park, 2007.

The new institutionalism

The idea of the new institutionalism came out of a paper by the political scientists James March and Johan Olsen written in 1984. In 'The new institutionalism: organizational factors in political life', March and Olsen noted a revival in political science studies looking at institutions themselves, and the significant influence they have on how political life operates and decisions are made and put into practice. The new institutionalism reacted against trends in political science in the 1960s and 1970s to downplay the importance of state institutions in how society operates, preferring a behavioural model that looks at the role of the individual. The state, in effect, was reduced to acting as a reflection of society, rather than actively shaping society into its own agenda. This approach brought with it a particular attention to the importance of shifting social class structures in the post-war period, with changing class structures subsequently being reflected in changing political landscapes and institutions.

Non-institutional interpretations of society emphasise the role of the individual. So the behaviour of markets, for example, is seen as the consequence of interlinking decisions by individuals to buy and sell according to their personal preferences. New institutionalism accepts some of this argument but suggests that rational choice by individual actors is not the only mechanism at play. In setting up institutions, which are of course composed of individuals, society creates a layer of regulations, rituals and ceremonies through which an institution acts. Procedures are established through which individuals in an institution react to certain events. These procedures become ritualised such that they become a purpose in their own right.

Inertia is a very powerful force here – 'we've always done things this way' – meaning that institutional responses do not change quickly to reflect the current state of society, as the non-institutional, behavioural argument would suggest. In turn, through the way in which they operate, those institutions can have a profound effect on shaping society. As March and Olsen (1984: 739) comment:

> Empirical observations seem to indicate that processes internal to political institutions, although possibly triggered by external events, affect the flow of history. Programs adopted as a simple compromise by a legislature become endowed with separate meaning and force by having an agency established to deal with them...

The impact of this kind of institutional structure on decision-making is outside the conventional understanding where people act to further their interests, depending on whether they have the power to do so and the constraints imposed by the existing legal framework.

New institutionalism therefore suggests that political preferences are shaped by education, indoctrination and experience – thus acknowledging the significant effect of institutional structures in shaping decision-making. This is, therefore, critical in terms of governance as decision-making is fundamentally shaped by the institutional structure of both state and non-state actors. In this light, the Thatcherite shift towards giving the non-state sector a much more prominent role in the delivery of services can be seen as even more dramatic as it overturned a series of institutional structures within the state that had been built up over several decades. New institutionalism also highlights the important effects of how governance arrangements are made, whether they be partnerships, regimes or networks, as these arrangements themselves are shaped by the principles under which they operate. The mission statement of a regeneration partnership, for example, can fundamentally shape the direction a redevelopment takes and how the actors within that partnership behave.

In theorising new institutionalism, March and Olsen (1984) suggested some three possible avenues of exploration. The first was the *policy martingale*. The idea of the martingale, a concept in probability theory, comes out of a gambling practice where a player with a one in two chance of winning a particular game doubled their stake with every losing bet – the idea being that they would eventually win back their money. March and Olsen use this idea to suggest that chance plays a major part in the policy process and that when a decision is made, this has a fundamental effect on the subsequent direction of policy-making. Essentially, one can conceive of the decision-making process as a series of forked branches with incremental decisions, influenced by random factors, pushing the process in one direction or another. Hence, even in nominally identical political systems, a particular set of circumstances will result in different sets of decisions being taken. This will in turn shape that decision-making process in particular ways and institutional responses will develop around this.

The second theoretical model March and Olsen identify is that of *institutional learning*. Where an institution feels that a decision taken has produced a successful outcome, they are more likely to reproduce that decision in the future. At the same time, institutions adapt their expectations of what comprises 'success' based on their past performance. Hence, if a particular regeneration project is judged to be a success, not only will an institution be more likely to do similar things in the future, it will also be more likely to redefine itself as an organisation which attempts to carry out projects of that nature. This can lead to a somewhat conservative decision-making process and make it very difficult to move in new directions.

The third of these models is that of the *garbage can*. This assumes that problems, possible solutions, decision-makers and opportunities flow through the policy system independent of each other and come together at random in a specific time and place. Hence, the way in which a decision is taken is entirely dependent on the chance coming together of particular decision-makers facing a particular problem with particular resources available to them. This model has certain attractions, though it is difficult to unpick the details of how this actually happens on the ground, especially where it is difficult to get access to the decision-makers sitting in particular meetings as decisions are made and acted upon.

When March and Olsen wrote their article back in 1984 their central message was that institutions themselves and the way they handled decision-making, had a significant impact on the process beyond the 'rational' choices of the decision-makers as individuals. At the time this was quite a significant break with behaviouralist models which emphasised individual choices. Today, however, the importance of institutions in the policy-making process is widely acknowledged, such that Pierson and Skocpol (2002: 706) can declare that 'we are all institutionalists now'.

Case study: the Greater London Authority

London, as Britain's biggest city, poses particular problems in terms of how it can be redeveloped. The formal institutions of local government in London have been in flux for some time, and as they have been reformed, so the opportunity has arisen to move the institutional aspects of London's governance in new directions. The Greater London Council (GLC), under the leadership of 'Red' Ken Livingstone in the early 1980s, had been vocal in its opposition to the policies of Prime Minister Margaret Thatcher. Thatcher responded by abolishing the GLC in 1986. Its powers and responsibilities were devolved downwards to the London boroughs, upwards to various central government departments and sideways to a raft of specialist agencies and **quangos**. The result was a bureaucratic mess which Tony Blair's New Labour government, elected in 1997, quickly acted to rectify, putting in place a new form of citywide government called the Greater London Authority (GLA). In the race to become Mayor of the new body, the old socialist stalwart Ken Livingstone ran as an independent, following a successful **Blairite** campaign to stop him from gaining the official Labour party nomination. In May 2000, Livingstone defeated Frank Dobson, Tony Blair's preferred choice, and was elected Mayor of London by a comfortable, if not overwhelming, margin. He was allowed to rejoin the Labour party in 2004 and won another four-year term, this time as the official party candidate, in May of that year.

Livingstone's capacity to irritate prime ministers aside, the institutional arrangements in which the GLA operates are very interesting. The Mayor is directly elected by the population of the London city region, producing by far the largest personal mandate of any elected politician in the UK. While the GLA comprises an elected assembly, its powers are in holding the Mayor to account rather than in playing an active role in policy-making – the Mayor alone having executive power. This combination of constitutional authority with the electoral mandate gives the Mayor considerable personal power within the remit of the GLA.

When defining that remit, however, there was a tension between whether the new London Authority would be based on the model of a local authority – as with the GLC before it – or something more akin to the devolved regional governments being set up in Scotland and Wales. Travers (2002: 781) has argued that the civil servants in the Scottish Office and Welsh Office had an incentive to ensure the new institutions being set up in Edinburgh and Cardiff had as many powers as possible as they would be transferred to administer the new bodies. Those Whitehall departments which shared responsibility for issues relating to London, however, would lose out on powers to administer these if a powerful London Assembly was established and thus there was more incentive to retain as many powers as possible within their own departments.

As a result, the GLA's formal powers are in some ways quite restricted. A series of existing bodies and funding streams were rolled together and the GLA was given control over four key agencies: Transport for London, the London Development Agency, the Metropolitan Police Authority, and the London Fire and Emergency Planning Authority. The GLA therefore has no formal role in education, for example, which remains in the hands of the individual boroughs.

This said, the Greater London Authority Act 1999, which established the GLA, defined the Authority as having a general purpose of promoting the social, economic and environmental development of London. The GLA therefore has a *general* responsibility to act in particular areas except where explicitly required not to, unlike ordinary local authorities where the roles they *can* play are strictly defined. Thus while the GLA is explicitly forbidden to reproduce activities being undertaken by statutory authorities such as health or education, it does have the authority to work in partnership with these bodies to promote the general well-being of the city.

The somewhat vaguely defined remit of the GLA, combined with the considerable personal authority which is invested in the Mayor, has meant that Livingstone has been able, on some levels, to shape the informal mechanisms and procedures of the new institution to his suiting. As Thornley et al. (2005) note, the agencies which were rolled into the GLA had already been working on strategic planning for the city and attempted to push these agendas into the formal London Plan which was produced by the GLA. Instead, the Plan was dominated by the personal, political, vision of the Mayor, who was able to stamp his authority on the newly coalescing institutional structures. Thornley et al. also argue that, by virtue of being involved in the process of setting up the GLA, business interests have managed to establish close links into GLA decision-making process. Indeed, a discourse of promoting 'competitiveness' is now central to the Mayor's remit.

The London Development Agency (LDA), which is overseen by the GLA, is the regional development agency (RDA) for the city. In common with the other RDAs, the LDA's flexibility has been significantly increased by the introduction of the so-called 'single pot' funding structure, which replaces the rather restrictive formula of several different funding streams for the projects it undertakes. This said, the governance of regeneration in London remains somewhat fractured. Indeed, the Local Strategic Partnership (LSP) mechanism was introduced after the foundation of the LDA and acts as a new tier of governance, matching local authority boundaries, and aiming to draw together local and national funding streams and priorities to tackle local social, economic and environmental problems. As originally envisaged, the LDA was primarily focused on tackling economic regeneration, with social aspects left to other agencies. In more recent years, the LDA's remit has drifted a little more into the social aspects of regeneration, for instance through the Opportunities Fund (2005–) which focuses on skills and capacity-building among deprived and socially excluded communities (LDA, 2007).

The LDA was given a significant boost by the successful bid to host the 2012 Olympic Games. In many ways, the LDA's role remains fairly conventional, however, focusing on land assembly and infrastructure. This role has not been uncontroversial, for instance proposing to build a large (if nominally temporary) car park on playing fields at Hackney Marshes (Samuel, 2005). The LDA also proposed compulsory purchase orders (CPOs) on 124 hectares in Stratford in order to make room for the Olympic park, dispossessing a number of existing businesses in the process (William, 2006). Approval for the CPOs was granted by the minister following a public inquiry. In the case of the London Olympics, of which both the GLA and central government are keen to make a success, there is a clear institutional framework for pushing ahead with these CPOs. This framework, combined with streamlined legal powers introduced in the Planning and Compulsory Purchase Act 2004, means that it is very difficult for local communities and businesses to stand in the way of a determined state.

Key points

- New institutionalism revived an earlier model of seeing institutions themselves as having a major role to play in how societies function – something which had been undermined by fashionable 'behaviouralist' approaches in the 1970s.

> *(Continued)*
>
> - 'We've always done it this way' – once procedures and systems are set up for carrying out a particular activity, institutions can be very reluctant to do things differently. This can have serious implications for the responses to rapidly changing circumstances.
> - The establishment of a new tier of government in London has created an opportunity to establish a new institutional structure, in turn creating a new way of 'doing' regeneration in the capital.
> - The importance placed on producing a successful London Olympics has led to the reshaping of certain powers for government institutions, in particular making compulsory purchase considerably easier, with huge implications for how urban regeneration schemes are carried out across England and Wales.

Community involvement

The example of the streamlined compulsory purchase procedure is interesting as it indicates that the state is frequently quite willing to crush local opposition to a particular regeneration scheme if that project is seen as a priority. This said, a discourse of community involvement is still central to the rhetoric of contemporary urban regeneration. Indeed, this can be seen as a reaction to the property-led, public–private partnership approach of the 1980s that produced schemes such as London Docklands. Through the 1990s, first through City Challenge and then the Single Regeneration Budget (SRB), central government funding for regeneration projects became tied to involving local community stakeholders.

Where a partnership or network of organisations is established to respond to a particular funding stream, as discussed above, not all of the actors within that network can be considered as equals. Where a community lacks a well-organised body to speak with one voice, or where competing community organisations have very different visions for the area, it is difficult to deliver a scheme with which most people are happy. Indeed, community voices can end up simply being marginalised, leaving behind a rather ambiguous discourse of communities being 'empowered' by the process (Atkinson, 1999).

The meaning of the word 'regeneration', as it has developed in the UK over the past couple of decades, is as an holistic concept – that is, on paper at least, regeneration should be tackling social, economic and environmental issues. It is interesting, however, that this holistic vision is starting to be broken down into two parts: *regeneration*, dealing with physical redevelopment, and *renewal*, dealing with social issues. Examining the regeneration of Nottingham as a case study, this extremely significant split in the discourse of regeneration is mirrored by the agencies which are leading in these two areas: physical regeneration is covered by the local urban regeneration company (URC), Nottingham Regeneration Limited, and social renewal by the Local Strategic Partnership (LSP) One Nottingham.

Nottingham Regeneration Limited is working on plans for three sites around the city core and proudly boasts that these represent £2.5bn of development opportunities. The plans for Southside (38 hectares), Eastside (56 hectares) and Waterside (100 hectares) include investment opportunities for new residential units, new offices, leisure facilities, retail and student accommodation. Urban regeneration companies in themselves tend not to have a great deal of resources but draw together key partners and funding streams. In the case of Nottingham Regeneration Limited, the main partners are the local authority, the regional development agency (East Midlands Development Agency), English Partnerships (the national regeneration agency), and the Greater Nottingham Partnership, a body set up in 1994 by the city and county councils which now manages the regional economic strategy for the area. Note the absence of any overtly community, voluntary or environmental stakeholders within the list of partners.

The local strategic partnership, One Nottingham, takes the lead on the community involvement aspects which seem to be missing from the institutional arrangement of Nottingham Regeneration Limited. The **discourse** of community 'renewal' comes from central government, with LSPs being established in response to a report by the Social Exclusion Unit (2001), *A New Commitment to Neighbourhood Renewal*. Resources are made available through the Neighbourhood Renewal Fund, set up in 2001, which distributed £1.875bn to the 88 most deprived local authority areas in its first five years.

Nottingham has an unenviable reputation for gun crime, ethnic tension and acute income inequality. The task for One Nottingham is to meet so-called floor targets for a range of socio-economic indicators – these are minimum standards for such things as crime, unemployment, income inequality, and so forth – and to try to narrow the gap between better-off and deprived communities. Nottingham was one of three LSPs singled out in 2006 for not producing convincing evidence of how they were acting to meet floor targets in deprived areas. Along with Hull and Birmingham, 10% of the Neighbourhood Renewal Fund resources allocated to these areas were withheld for four months. One Nottingham subsequently found itself with a target of reducing overall crime in the city by 26% in two years by expanding neighbourhood policing and forming a specialist Drugs Squad (CLG, 2006a).

As should be apparent from this example, even in the 'renewal' sector, there is still a great deal of top-down imposition from central government upon which funding streams are dependent. Indeed, Diamond (2004) has argued that LSP managers themselves perceive the community as a weak, 'dependent' partner in the process as those managers strive to meet performance targets set down centrally. It is perhaps somewhat troubling that this new tier of the state apparatus, focused on the 'community' yet completely lacking a democratic mandate, struggles to form genuine partnerships with community groups while at the same time legitimising the idea that a physical regeneration process can somehow be institutionally separated from the social issues.

Academics examining the involvement of local communities in decision-making have created typologies, or ladders, with different levels of community participation. The idea of a ladder, was devised by Sherry Arnstein (1969), a planner, who perceived three tiers of participation (Figure 3.2). Non-participation occurs where community voices are manipulated or are given a forum to discuss issues ('therapy') without actually having any impact on the process. More tokenistic involvement could include simply informing the community

	Level of participation

Degrees of Citizen Power
  Citizen Control
  Delegated Power
  Partnership
 	
Degrees of Tokenism
  Placation
  Consultation
  Informing
 	
Non-participation
  Therapy
  Manipulation
	↓

Figure 3.2   Sherry Arnstein's ladder of citizen participation in decision-making. Nearly 40 years on, participation in UK urban regeneration seems all too often to be located towards the tokenistic end (derived from Arnstein, 1969)

of the decisions which were being made and the issues at stake, or engaging in a formal consultation process which would not necessarily result in any change to the process. Placation, though still tokenistic, would involve giving hand-picked members of the community seats on various decision-making boards. More realistic degrees of citizen power come through making communities a partner in decision-making or actually delegating power down to the communities to be the majority decision-maker over particular issues. At the top of Arnstein's ladder was some form of citizen control, for example a neighbourhood corporation with managerial control over, for instance, a school.

The accusation levelled at regeneration projects is that the degree of citizen participation generally comes rather low down on Arnstein's ladder. While the LSPs have significant community input, the target-driven culture for their management means that they remain somewhat top-down organisations. This said, the rhetoric of involvement and participation is now deeply embedded in regeneration programmes, even if the reality is somewhat less encouraging. One of the most significant sources of funds for regeneration was the Single Regeneration Budget, which has since been replaced by the Single Programme ('single pot'). Both SRB (which ran from 1994 to 2001) and single pot require community organisations to be included within partnerships bidding for funds. The bureaucratic complexity of the bidding process effectively restricts the leadership of bids to the local state and well-funded third-sector organisations, meaning that the involvement of the actual community in community projects has the potential to be more apparent than real. The system therefore brings with it the potential for a cynical local authority to simply use a community group as a purely nominal partner in order to access a funding stream. At the same time, the very notion of a single 'community' voice is somewhat ridiculous, making it difficult to fit 'the community' neatly as an actor into a process with private and public sector partners (see Box 3.2)

Box 3.2   Elephant Links Community Forum (North, 2003)

A regeneration programme proposed by Southwark Council in the late 1990s combined a Public Private Partnership undertaking physical regeneration and a £25m seven-year SRB application looking at environmental improvements and tackling social exclusion. Although a Stakeholders' Forum was established to help shape the SRB bid, much of the work was done by the Council and presented to the forum late in the process, causing some community representatives to threaten to withdraw their support.

The SRB funding was granted in 1999 and a Community Forum, eventually including 63 local tenants' and residents' groups, was established to be one of the partners. The Forum had some successes, securing resources so that it had a more professional organisational capacity and gaining a voice on the physical regeneration programme which was running in parallel to the SRB process. There were, however, conflicts within the Forum and even walk-outs by key groups, and the Council stepped in to counter the 'unruly' behaviour of the Forum, withdrawing its resources. At the same time, the large physical redevelopment programme was cancelled in 2002, with the Council looking to break up the project into more manageable pieces and attempting to bypass the involvement of the Community Forum. The Council even went to Court to require the Forum to return key documents relating to the original project.

Existing SRB projects continued to operate for their seven-year cycle even after the scheme was phased out in 2001. The replacement single pot scheme is administered by the regional development agencies (RDAs) – a crucial change in institutional arrangements. Although the single pot represents a considerable simplification of the multiple funding streams that existed before, in theory making 'holistic' projects more feasible, in practice the RDAs remain primarily economically focused. Indeed, a report by Urban Forum (2004) commented that voluntary and community organisations, which had not always had a good relationship with SRB, were severely troubled by single pot. Partly this was about having a new set of forms to fill in, but it was also a fear that community/social programmes would be marginalised in favour of economic ones, although those groups involved in social enterprise activities did see single pot as an opportunity. Again, the fear is that the bureaucratic arrangements weaken the position of community groups in the governance of regeneration. Similarly, the institutional switch to using RDAs has the potential to skew the programme simply because of the nature of the institution which has been handed the job.

Key points

- While there is an emphasis on getting communities involved with regeneration projects, in practice the community is often the weakest partner, particularly where there is not a well-organised community body with a clear agenda (very difficult to produce in practice).

(Continued)

- Increasingly, the community aspects of regeneration are being hived off into a discourse of 'renewal', involving separate bodies such as Local Strategic Partnerships. This breaks down the idea that regeneration is an holistic approach which does not simply focus on bricks and mortar.
- There has been considerable difficulty in moving community involvement into the more meaningful 'participation' areas of Arnstein's ladder and away from merely keeping communities informed of what is happening.
- The institutional switch to 'single pot' funding controlled by the regional development agencies is of some concern to community organisations, which are worried that this will skew projects receiving funding to those with more 'economic' concerns.

Regional regeneration and the European Union

Since 1975 the European Union (at that time named the European Economic Community) has been providing funds to sub-national regions of the member states under the European Regional Development Fund (ERDF). Objective 1 funding tackles structural change in particularly economically deprived areas. Objective 2 funding tackles specific socio-economic change, particularly relating to industrial and rural areas. Where regions are deemed to qualify for these funding streams, quite considerable extra resources are available outside the ordinary local and national state sources. The availability of EU funding, bypassing national governments, forms part of Jessop's (1994) argument about the hollowing out of the central state and the process of glocalisation discussed above.

The semi-autonomous regional governments in Scotland and Wales carry the responsibility for coordinating bids by local authorities and others to the ERDF. In England, these bids are coordinated by the nine Government Offices for the Regions. The Government Offices are staffed by central government civil servants and are directly accountable to Whitehall. Central government thus retains a high degree of control over ERDF bids, which somewhat undermines the notion of glocalisation. Bids can be made by private and voluntary sector organisations as well as local authorities and regional development agencies (RDAs). Regions have to agree a Single Programming Document (SPD) of priorities with the EU, which are designed to map on to the strategy of the relevant RDA. Again, given the dependence of the RDAs on central government funding and the skew this gives to their priorities, this reinforces the influence of Westminster over ERDF bids.

The Government Office administrative region for the north east of England contains large areas which fall under the Objective 2 criteria for ERDF. In the funding round 2000–06, over £500m was secured for the region. This included a wide variety of projects from Newcastle's Seven Stories Centre for the Children's Book, which opened in 2006, to the Seaham Regeneration Scheme, which created a new road and footpath infrastructure as well as brownfield **remediation** (CLG, 2006b).

In Wales, since 2000, bids to the ERDF have been coordinated by the Welsh European Funding Office (WEFO). Taking account of all European funding streams, Wales secured

£1.6bn in the period 2000–07, which is a significant sum of money for a country of around 3 million people. This funding reflects the significant structural problems which continue to be faced by the Welsh economy following the collapse of the mining and industrial sectors in the 1980s.

Cardiff qualifies for Objective 2 funding and brought in some £7.3m in 2000–07. This money is spread quite thinly, however, across some 37 projects, the smallest of which was just £1,800 for a feasibility study looking at job creation in the Adamsdown area (WEFO, 2007). One of the most significant urban regeneration interventions in Cardiff was the cleaning up of the waterfront area, which is now the home of the Welsh Assembly Building (£67m, designed by Richard Rogers) and the Welsh Millennium Centre (£106m, designed by Percy Thomas). While the Cardiff Bay project did attract around £7m from the ERDF, this was dwarfed by the £496m in UK government funding via an urban development corporation and the £1.14bn brought in from the private sector (Cardiff Harbour Authority, 2007). While EU funding can be extremely important, therefore, it is rarely the sole source of revenue on a project and often only pays for quite small elements in a much larger programme.

Key points

- The European Union has become increasingly important as a source of funding for regeneration projects, although in practice projects require a portfolio of different funding sources.
- The importance of the EU for funding adds some weight to the 'glocalisation' argument that national governments are increasingly being bypassed, although in practice in England and Wales overall control over bids to the EU are regulated by central government.

Further Reading

Governance is a diverse and complex topic. Newman (2001) provides a good introduction to the development of governance in the UK since the election of the Labour government in 1997. Davies' (2001) exploration of urban regeneration and regimes provides a good critique of Rhodes' work on network governance. March and Olsen (2005) have written a good summary of the development of the new institutionalism in the 20 years since they devised the term. Imrie and Raco's (2003) edited collection provides excellent material on community involvement. Marshall's (2005) article provides a good critical examination of the impact that European funding programmes are having on how local authorities organise themselves.

Davies, J. (2001) *Partnerships and Regimes: The Politics of Urban Regeneration in the UK* (Ashgate, Aldershot).
March, J. and Olsen, J. (2005) *Elaborating the 'New Institutionalism'*. Arena Centre for European Studies Working Paper No.11, University of Oslo, Oslo.

Imrie, R. and Raco, M. (2003) *Urban Renaissance? New Labour, Community and Urban Policy* (Policy Press, Bristol).

Marshall, A. (2005) 'Europeanization at the urban level: local actors, institutions and the dynamics of multi-level interaction', *Journal of European Public Policy,* 12(4): 668–686.

Newman, J. (2001) *Modernising Governance: New Labour, Policy and Society* (SAGE, London).

4 The Competitive City

Overview

This chapter identifies key approaches to the economic regeneration of cities attempting to recover from deindustrialisation, and explores the outcome of current thinking through a series of case studies. The chapter is structured as follows:

- *Deindustrialisation and the competitive city*: discusses the decline of British cities and the subsequent competition to attract investment in the new economy, setting urban regeneration within this context.
- *Funding economic regeneration*: sets out the key policies and funding mechanisms used to drive economic regeneration.
- *Regenerating cities in practice*: uses a series of case studies to demonstrate and evaluate the economic success of urban regeneration.
- *The entrepreneurial city*: explores how cities are responding to the knowledge economy and outlines policies to encourage enterprise.

Deindustrialisation and the competitive city

Deindustrialisation and globalisation

While the symptoms of urban failure are poverty, crime and dereliction, the underlying causes tend to be economic. In order to understand why urban regeneration is needed, it is first necessary to appreciate the economic history of British cities. The modern British city emerged in the industrial era of the eighteenth and nineteenth centuries, and the waning of the British manufacturing sector in the second half of the twentieth century has been the main cause of urban decline. Industry in north America and Europe was undercut by

Table 4.1 Population change by percentage in British cities during the twentieth century (Hart and Johnston, 2000; Office for National Statistics, 2002)

City	1901–51	1951–61	1961–71	1971–81	1981–91	1991–01
Birmingham	+49.1	+1.9	−7.2	−8.3	−5.6	+0.7
Glasgow	+24.9	−2.9	−13.8	−22	−14.6	−4.7
Leeds	+19.3	+2.5	+3.6	−4.6	−3.8	+5.1
Liverpool	+10.9	−5.5	−18.2	−16.4	−10.4	−2.8
London	+25.9	−2.2	−6.8	−9.9	−4.5	+10.5
Manchester	+8.3	−5.9	−17.9	−17.5	−8.8	−2.1
Newcastle	+26.1	−2.3	−9.9	−9.9	−5.5	−0.1
Sheffield	+23.0	+0.4	−6.1	−6.1	−6.5	+2.5
UK	+32.1	+5.0	+5.3	+0.6	+0.02	+2.3

cheaper and often superior goods produced by emerging countries in the Far East, such as Japan, Korea and Taiwan, with lower labour costs and often more efficient production processes (Dicken, 2003). This is a trend that has continued apace with China's emergence as the major global manufacturing power. The effect of deindustrialisation on British cities was devastating. From the closure of steel plants in Sheffield in the 1980s and the rapid decline of motor industry in the West Midlands to the rationalisation of chemical plants on Teesside and the loss of shipbuilding from Newcastle, high levels of unemployment became synonymous with urban life. Between 1971 and 1981 Britain's cities lost 34.5% of their manufacturing industry, equating to almost 1 million jobs. Birmingham lost nearly 60,000 jobs between 1981 and 1992, which represented a decrease of over 11%, compared to a regional average loss of just 0.7% and a national *gain* of 5% (Duffy, 1995). At the same time, new businesses preferred to locate outside major cities. There was more room to build larger integrated production lines associated with sophisticated production processes. This was in stark contrast to the urban areas vacated by industry that were characterised by antiquated buildings, obsolete **infrastructure** and high levels of pollution. The development of the motorway network also meant that components and products could be moved around easily without needing centralised rail stations and storage depots.

The deindustrialisation of cities in the UK was accompanied by an equally dramatic loss of population. From the 1930s onwards, there was a selective migration of the wealthier and more educated middle and upper classes out of Britain's industrial cities to rural locations, suburbs and new towns. As Table 4.1 shows, the trend of urban population loss in the 1960s, 1970s and 1980s was uniform across all major UK cities, with Liverpool, Manchester and Glasgow hit particularly heavily. The upturn between 1991 and 2001 shown in the final column is a direct result of urban regeneration policies.

This selective out-migration tended to leave poorer, less skilled populations in the city. The negative effects of this movement only became acutely felt in the 1970s and 1980s when industrial decline accelerated. Further, new retail and leisure developments tended to follow the geography of consumer spending power and also abandoned the city centre for out-of-town locations near to major road networks.

At the same time as manufacturing industry shifted to the emerging Tiger economies of south east Asia, the western economies became increasingly reliant upon the service and finance sectors. This so-called 'new economy' is based on services, communication, media and biotechnologies, and tends to be characterised by information and knowledge-intensive activities. Some commentators believed that the forces of globalisation and new economy would spell the end for cities, suggesting that in a world where money and information could be sent instantaneously and electronically from any location to any other there would cease to be a need for concentrations of people and industry. We were offered a future in which everyone could work at home (or in the Bahamas, or, indeed, anywhere they wanted), but this prediction proved to be wide of the mark. Rather than cities becoming obsolete, the new economic era heralded the advent of greater metropolitan dominance by an elite group of *global cities*, such as Tokyo, New York and London.

The urban theorist Saskia Sassen (1994) argues that cities remain necessary because the work of global integration has to be done *somewhere*. Analysts tend to distinguish between codified knowledge that is widely available, for example through the internet, and tacit knowledge that is generally guarded (Lever, 2002). While codified knowledge can furnish information concerning markets, services and processes that are valuable, it does not necessarily confer a *competitive* advantage, as technically everyone can access it. Knowledge concerning investment and innovation is often only passed on through face-to-face interactions, in order to ensure confidentiality and the correct interpretation of information. This knowledge can confer a competitive advantage, and tends to be associated with higher-level financial and technological industries. While the 'global cities' hypothesis has been criticised, it is generally accepted that cities which provide a place in which these face-to-face interactions and innovations can easily occur tend to enjoy a considerable advantage. People have to sell and consume high-level goods and be educated and trained somewhere.

Global cities provide command points for controlling the new economy and thus have much more in common with each other than they do with other cities in their own country. This creates massive imbalances in wealth between cities (and parts of cities) that get a share of the global pie and those that do not. The glaring concentration of wealth in the south east of England around London is testament to this phenomenon. France is also characterised by a growing economic gap between Paris and the rest of the country, referred to in the popular press by the phrase 'Paris et le desert Français'. Indeed, while traditionally policy has tended to make noises about reducing intra-national imbalances, some groups are now calling for global cities like London to receive preferential treatment as they contribute so much to the national economies of which they are a part.

The new economy and the competitive city

Globalisation, the new economy and the emergence of global cities frames economic regeneration policy in the UK. As manufacturing industry shifted its focus to overseas locations, cities began to move their attention towards encouraging an alternative basis for their development in the new economy. Within the context of deindustrialisation and depopulation, the main goal of regeneration is to generate employment, preferably in the form of middle- to high-income jobs in the service sector. While relatively footloose in terms of material resource demands, service industries tend to cluster in areas that provide attractive living and business environments (we will return to the idea of clusters later). Cities seeking to stimulate economic growth compete with one another to attract such industries, whereby they must 'sell' themselves as desirable locations within the new economy.

The idea of 'competition' is commonly used to describe this situation, and in 1997 the UK government funded a major five-year research programme called 'CITIES: Competitiveness and Cohesion' to investigate these processes further. The programme identified two ways in which the competitiveness of cities is commonly understood (Begg, 2002). First, cities can be viewed as competitive if they offer a place in which it is cheap to do business; secondly, cities can be viewed as competitive if they have a mix of attributes that is more attractive to key industries. Because the idea of urban competitiveness is fairly vague, it is hard to measure. The research identified three conceptual approaches to measurement: economic outputs, such as income, unemployment, growth; intermediate measures of success, such as visitor numbers and student population; and more general indicators of quality of life. While the idea of competitive cities is undoubtedly here to stay, it is worth noting that some economists have argued that only firms compete, and thus it makes no sense to apply the term to territorial units such as cities (Krugman, 1996).

Within the new economy a range of attributes are desirable. In order to attract business and visitors cities seek to establish cultural pursuits such as theatre and arts, shopping facilities, sports and conference facilities and an attractive living environment. The physical built environment constitutes the 'hard' assets that drive economic regeneration. Modern offices are required in order to attract white-collar employers. Creating these environments goes hand in hand with large-scale infrastructure planning. An effective modern transport system is necessary to allow access to attractions and reduce the negative effects of congestion and air pollution. Information communications technology (ICT) infrastructure is increasingly critical, providing broadband to private homes and schools, and wireless broadband in public spaces. Regeneration projects need to be planned, financed and executed in order to ensure that new facilities are located in the right place with the right mix of land uses. Other desirable characteristics are less easily created, such as the presence of countryside and attractive villages nearby, in which potential managers may live (Duffy, 1995).

Professor of Regional Economic Development at Carnegie Mellon University, Richard Florida (2002), has taken these arguments further, claiming that only those places able to attract what he calls 'the creative classes' will prosper in the new economy. He defines this category as a select group of people who are employed in 'science and engineering, architecture and design, education, arts, music and entertainment, whose economic function is to create new ideas, new technology and/or new creative content' (Florida, 2002: 8). This emergent class is distinct from the service sector class because they are primarily paid to create rather than to execute orders and are typified by very high levels of education and human capital. In the vanguard of the new economy, then, Florida identifies the emergence of a 'knowledge economy', in which the most valuable asset is information and know-how. Florida claims that they are increasingly important drivers of economic growth, as they tend to earn twice as much as average workers and work in key growth industries such as IT and biotechnology. He argues that this class of people value specific lifestyles that incorporate individuality, self-expression and openness to difference, and that these moral norms and desires are seen as inseparable from creative work. The creative class seeks what he calls an 'experiential lifestyle', which offers a range of creative experiences that complement their less constrained working arrangements. The key message for cities is that in order to attract these people they need to become capable of delivering these kinds of lifestyle.

As a result, cities have tried to create attractive living environments that offer high-specification housing alongside recreational and leisure facilities. Cities have traditionally been diverse places, home to the widest ethnic and demographic cross-sections of society, and urban planners and marketers play on these legacies. Mixed land uses are complemented by efforts to impart historical identity to regenerated areas and provide individual lifestyle environments such as loft-style apartments with home office space, and unique retail spaces. But while cities are undoubtedly at an advantage in terms of diversity, this does not necessarily mean that people will be more tolerant of difference. Florida (2002: 256) cross-referenced the 'Gay Index', which measures the total gay population in US cities as a proxy measure for liberalism and tolerance, with levels of innovation and high-tech industry. He found a very high correlation between gay population (i.e. tolerance) and concentrations of high-tech industry and innovation. These findings generated significant interest, as 'soft' cultural factors such as tolerance and diversity were suddenly shown to be critical in affecting the economic fortunes of a city. While much of Florida's thinking about diversity and experiential lifestyles informs urban regeneration policy in the UK (albeit often indirectly), the task of making existing urban inhabitants more liberal and tolerant towards the creative classes and those groups that attract them may be harder to address.

Some of the demands of the knowledge economy are more direct, but no less obviously satisfied. Florida argues that cities with a more educated population will gradually become more prosperous over time, while those with less educated populations will fall further behind. Creating a skilled population that is able to work within a knowledge-based sector requires education initiatives at school, university and adult levels. This involves encouraging more people to stay in education for longer and retraining unemployed industrial workers. Within the knowledge economy, higher education institutions like universities and research institutes act as centres for education and innovation. Harder questions related to education involve how to encourage entrepreneurship and business start-ups among the population of a city.

Three points merit further attention. First, the attributes identified by Florida are not only very different from those demanded by manufacturing industry, but are almost *antithetical*. Cities like Sheffield, which have suffered from deindustrialisation, are typified by a large pool of unskilled labour and a deteriorating urban environment that may be visually unappealing and socially undesirable. The image of British cities, particularly the inner cities, in the 1980s was incredibly bad – these were not places that people would move to out of choice. This was obviously a major problem when trying to attract footloose knowledge-intensive industries. Because cities require a complete face-lift in order to make them attractive destinations, it is hard to disentangle purely economic measures from more general regeneration measures to encourage cultural amenities and improve the environment.

Secondly, the attraction of a city will vary for different types of business. These different dimensions of competitiveness mean that different cities will be competing in different markets. So, for example, London courts international financial business and thus competes with New York and Tokyo, while Cambridge seeks to attract high-tech ICT industries and thus competes with Reading and Oxford. The major regenerating regional cities of the UK (such as Manchester, Glasgow and Leeds) are competing against one another primarily for business within the domestic market, while simultaneously trying to establish themselves as European and global players. The group of global cities constitute an exclusive club, but the potential rewards of breaking into this group means that these cities expend vast efforts attempting to do so. While the new economy tends to be represented as an homogeneous and sweeping phenomenon, the place of cities within the global context is necessarily more differentiated than the term would suggest, responding to different economic opportunities and niches. This allows space for policies to be tailored to the unique characteristics of individual cities. The examples in this chapter and Chapter 6 on design and culture show how regenerating cities attempt to establish a specific place-identity.

Thirdly, Florida's notion of diversity has been criticised for failing to describe the distribution of creative economies. For example, when he tested his indices of diversity in the UK, Manchester emerged as the most creative city, followed by Leicester. As Montgomery (2005) has noted, while Manchester is indeed a centre for creativity, it is outstripped by London's creative economy on every count – from number of employees and companies to overall economic output. Similarly, Leicester scores highly on account of its ethnic diversity, but is far from being a hotbed of innovation and creativity. Florida's three Ts of talent, tolerance and technology should not be seen as a one-size fits all solution, but as part of a more complex suite of factors that come together to allow a city to regain its economic prosperity.

Urban regeneration to the rescue

So far this chapter has established that the problems facing UK cities are related to their general economic decline over the course of the twentieth century, and that the goal of regeneration is to enable cities to compete within the new economy. This is a critical task. Cities are vital to the prosperity of the UK as a whole, which explains in part the political priority afforded to their regeneration. In 2004, London contributed 17.9% of the UK's overall GDP, while housing 12.4% of the national population (Office for National Statistics, 2007). Similarly, a recent government paper (HM Treasury, 2001) highlighted the strong links between the economic performance of cities and their regions – for example, Manchester generates 42% of the GDP of the north west region. Jane Jacobs (1985) argues that as engines for regional innovation and growth, cities are more important economic units of study than nations, and policy-makers are now talking about the 'city-region' as a focus of concern.

Economic regeneration can be split into external measures, designed to attract investment in from outside, and internal measures, designed to stimulate local enterprise. Externally, the regeneration paradigm of mixed-use planning is informed by the need to attract economically viable land uses – office space located in the same development as living and recreational space, and so forth. Taking this a step further, the social and environmental dimensions of regeneration are often justified as a necessary step in encouraging economic regeneration. When we look at the goals of the Urban White Paper (DETR, 2000) concerning the creation of high-quality built environments, the underlying aim is to attract *people* back into the city. People who work, consume and run businesses that generate jobs and wealth. In this way, the economic imperative underpins the entire urban regeneration agenda.

In addition to encouraging cities to compete externally, policy-makers are realising the internal importance of cities as centres of innovation and enterprise within regional economies. Economic regeneration thus deploys local economic policies and initiatives to unleash forces of local enterprise and innovation within the city. The remainder of the chapter considers the key external and internal policy initiatives that have been developed to stimulate economic regeneration, using case studies to reflect upon their successes and failures.

Key points

- In the twentieth century UK cities were hit by the negative consequences of deindustrialisation, including chronic depopulation and widespread dereliction.
- The overriding goal of urban regeneration is to revive economic growth by attracting investment and people back into cities.
- Within the context of globalisation, this involves courting the so-called 'new economy', which is typified by knowledge-intensive industries such as the service and ICT sectors.
- Regeneration is often thought of in terms of competition between cities.

Funding economic regeneration

At the most basic level, the challenge facing post-industrial cities has been to deal with the problems of inner-city dereliction. In classic urban land rent models, property values increase with proximity to the city centre. Widespread dereliction caused by people, retail and factories leaving the city can lead to the reversal of this relationship, with rent becoming cheaper in more central locations, causing market failure. As a result of cheaper land being available, low-density land uses, such as car sales, move in, which has the effect of progressively dispersing the city centre. Lower densities of land use and population make it impossible to maintain an efficient urban infrastructure. Without any core services and facilities, the city centre will gradually die. The financial demands of addressing these physical, social and economic problems are massive, making funding a key issue within urban regeneration.

Central government involvement was essential in the face of these challenges, and the Urban White Paper of 1977, *Policy for the Inner Cities* (Department of the Environment, 1977), kick-started urban regeneration, increasing government funding from £30m to £125m per year (Noon et al., 2000). Land-use planning is probably the most important form of local regulatory intervention in the economy, as it has the power to alter the physical landscape of cities. After their election in 1979, Margaret Thatcher's Conservative government focused on property-led regeneration as the means to re-establish higher rents in city centres. Urban development corporations (UDCs) were set up in 1980 to encourage the redevelopment of land in depressed urban areas. A key activity involved assembling land packages from diverse private owners and improving infrastructure to attract investment. Public funds would be focused on rejuvenating the physical infrastructure of an area, which would then attract private developers. It was intended that government funding would be supplemented by subsequent profits from land sales as an area became more desirable (known as the multiplier effect), but these returns were decimated by the property market crash that took place in late 1980s.

Early regeneration tended to be coordinated at the national level and driven at the local level by organisations which were created to function independently from city councils. This partly reflected the inherent mistrust of public institutions held by Thatcher's right-wing government, but also represented the recognition that there were serious institutional constraints on urban economic growth. City leaders tended to be more concerned with the services that are provided to the residents, such as refuse collection and council tax, than with responding to the latest economic trends in the global market. In order to achieve comprehensive change, redevelopment efforts were focused on specific areas, an approach sometimes termed Area Based Initiatives (ABI). The logic was that limited funds would be more effective if they were concentrated in relatively small areas, rather than being spread thinly across entire cities. This would then cause a 'trickle-down' effect, as benefits accruing in the regenerated area spread out to the areas surrounding it.

While achieving successes in some areas (for example, the London Docklands attracted £2bn of funding in 1989), the property-led regeneration policies of the Conservative government have been criticised on a number of counts (Imrie and Thomas, 1993):

- for failing to address social and environmental problems;
- for being driven from the top down and implemented by government **quangos** that failed to respond to local community needs; and
- for using public money to subsidise infrastructure for private developers.

An early evaluation (Robson et al., 1994) of the economic impacts of urban regeneration between 1983 and 1991 suggested that increased expenditure in an area led to some reduction in unemployment and the retention of 25–34 year olds. But out of 57 areas, only 18 had positive effects and 21 had poor outcomes. The report argued that in many schemes the focus on private property investment had alienated local communities and failed to utilise the skills and commitment of local people. Projects needed to be more focused on local needs and inclusion through adopting a partnership approach. The failure of private developers to rejuvenate city centres indicated the need for a more comprehensive approach to regeneration, capable of generating better value for money for the public purse.

City Challenge and the New Localism

The 1990s saw a major shift in urban regeneration, away from grants and quangos towards integrated regeneration projects that were controlled by local councils. The first of these approaches was City Challenge, introduced in 1991. City Challenge was intended to be different in terms of its values, organisation, scope and delivery, and it is worth considering its main tenets as they set the tone for subsequent urban regeneration funding policy:

- *It adopted a comprehensive and strategic approach* that sought to address physical regeneration alongside economic and social problems.
- *It was limited to a five-year period* with each City Challenge adopting the same timescale with the same level of grant allocated in equal annual instalments.

- *It was competitive*, allocating £37.5m each to 31 Urban Programme authorities over five years on the basis of two competitions.
- *Bids were put together by cross-sectoral partnerships* of public, private and voluntary bodies.
- *It emphasised local implementation*, allowing local authorities to select areas and draw up plans to allow sensitivity to the demands of local circumstances.
- *Flagship developments were strategically targeted at specific areas* in order to kick-start further development.

City Challenge ushered in an era of '**New Localism**' within UK urban regeneration, whereby regeneration projects became driven by individual cities, but in a highly managerialist, competitive and corporatist manner (Stewart, 1994). Successful bids had to identify measurable outputs and draw up delivery plans with their partners in order to compete with rival bids. While the competition format bred success within some local authorities, it penalised others that were unable to develop plans in such a comprehensive format and demonstrate an ability to deliver. For example, Bristol failed to secure City Challenge resource in either funding round (Malpass, 1994), and the neoliberal agenda of competitiveness has been criticised for divisiveness. Despite this, the logic of competitiveness has remained a dominant theme in regeneration funding, and councils today put together bids to attract national or international facilities. Recent examples of this process include the national bidding process for the new Wembley, and the location of 17 new casinos in 2007.

The shift to regional development

The ABI approach to regeneration was extended in 1994 with the introduction of the Single Regeneration Budget (SRB). Government evaluations of urban regeneration funding found that certain policies overlapped, and that this was hindering efficiency and cost-effectiveness. The SRB brought together a number of programmes from several government departments, simplifying and streamlining the assistance available for regeneration into one funding stream. A review of the first three years of SRB highlighted problems with governance, resources and policy (Hall and Nevin, 1999). As with City Challenge, the competitive allocation of resources created winners and losers in the game for regeneration funding. If anything, SRB exacerbated this problem by forming a single resource, which meant that partnerships had to meet the stated criteria or else they ran the risk of receiving no resources. For example, over the first three rounds of SRB funding Leicester attracted £12m while Newcastle received £109m, despite the similarity between the two cities in terms of socio-economic deprivation. The amount of money made available under the Conservative government through SRB also fell by some 33% between 1994 and 1997, causing a continued focus on economic regeneration at the expense of social and environmental issues.

In 1997 the newly created Government Offices for the Regions were given the role of administering this budget. Responsibility was subsequently transferred to the regional development agencies (RDAs) when SRB was replaced by the Single Programme ('single pot') funding scheme in 2001. By administering the funds regionally, it was hoped that the process would be more flexible and responsive. The regional agenda has a strong presence at the moment, and the current governance of economic regeneration is primarily undertaken at the regional level. Politicians talk about a 'Europe of the regions' and, on the back of the successes of Scottish and Welsh devolution, the present UK government is trying to create politically active regions in England.

Under the Regional Development Agencies Act 1998, each agency has five statutory purposes:

1   To further economic development and regeneration.
2   To promote business efficiency, investment and competitiveness.
3   To promote employment.
4   To enhance development and application of skill relevant to employment.
5   To contribute to sustainable development.

The RDAs are expected to take a proactive role in establishing regional competitiveness by encouraging inward investment and working with regional partners, and they play a leading role in identifying and funding urban regeneration projects. These goals mirror the economic aims of the single-pot bidding guidance to improve the employment prospects, education and skills of local people, and support and promote growth in local economies and businesses, but only provide for environmental and social considerations through their final role to contribute to sustainable development.

City Challenge and SRB represented a commitment to increasing local participation in urban regeneration, and the further devolution of the single-pot administration to the regional development agencies can be seen as a continuation of this trend. But while the regional development agencies are publicly accountable, they are not publicly elected. When the regional development agencies were given control of single-pot funding it was envisaged that the newly established and democratically elected regional assemblies would provide a supervisory function over how the money was allocated. With the exception of London, the regional assemblies have failed to become established as democratic entities, and single-pot funding is almost solely controlled by the regional development agencies. As a result the specific economic brief of the regional development agencies has lent the regeneration funding process a pronounced economic bias.

European funds

Although European funding is primarily allocated at the regional level, it has played a major role in urban regeneration because many depressed regions of the UK contain extensive metropolitan areas. The original European Economic Community stressed the need to reduce inequalities between the regions of its various members, setting up the European Regional Development Fund (ERDF) in 1975 to provide investment in socially and economically challenged areas of Europe. While the fund initially targeted predominantly agricultural economies such as Greece and Ireland, deindustrialising regions in the more developed nations began to receive funds in the 1980s. For example, Birmingham received £260m between 1985 and 1994 (Duffy, 1995), which was then matched again by national and local government funding. Through the ERDF, £5bn has been invested in the regeneration of over 300 English communities since 1975. Currently, the Structural and Cohesion Funds are divided into three separate funds: the ERDF, the European Social Fund (ESF) and the Cohesion Fund. These are used to meet the three objectives of cohesion and regional policy: convergence, regional competitiveness and employment, and European territorial cooperation. The EU budget for these funds runs on a six-year cycle, and approximately £3.5bn of funding was made available to the English regions between 2000 and 2006, with roughly the same amount going to Wales, Scotland and Northern Ireland.

The money is split into Objective 1 and Objective 2 funding. Objective 1 funding promotes the development and structural adjustment of highly specific areas whose economies are in decline. Objective 2 funding aims to assist regions whose economies are facing structural difficulties, whether they are industrial, rural or urban. The amounts given to each region differs according to need, which is generally determined by their productivity relative to the EU average. Thus north west England received £511m of Objective 2 funding, while south east England received only £23m. Objective 1 funding is more spatially focused, and only three English regions qualify to receive it: the South West (Cornwall £190m), Yorkshire and Humberside (South Yorkshire £497m) and the North West (Merseyside £565m). West Wales and the Valleys, Northern Ireland and the Highlands and Islands of Scotland also receive considerable Objective 1 funding. The UK will continue to receive substantial Structural Funds, amounting to just under £6bn for the period 2007–13, although this is less than the previous period 2000–06, as funds are being diverted to the various Eastern European countries that have joined the EU.

EU legislation also constrains UK government funding for regional development, only allowing direct state aid to be given to areas that qualify for 'assisted area' status. These areas are defined by employment rate, adult skills, incapacity benefit claimants and manufacturing share of employment. Regional aid consists of aid for investment granted to large companies or, in certain limited circumstances, operating aid that is targeted at specific regions to redress regional disparities. In the UK, the main forms of regional aid are channelled through discretionary grant schemes:

- *Selective Finance for Investment in England* (SFIE) – funds new investment projects that lead to long-term improvements in productivity, skills and employment.
- *Regional Selective Assistance* (RSA) – administered by RSA Scotland, part of the Scottish Executive, aimed at encouraging new investment projects, strengthening existing employment and new job creation.
- *RSA Cymru Wales* (Regional Selective Assistance) – delivered by the Welsh Assembly government to help support new commercially viable capital investment projects that create or safeguard permanent jobs.

Importantly, funding is only allowed in areas that qualify for 'assisted area' status. Figure 4.1 shows the coverage of status for the UK for the period 2007–12. Aid to promote the economic development of areas where the standard of living is abnormally low or where there is serious underemployment is covered under Article 87(3)(a) of the EC treaty on state aid rules, while aid to facilitate the development of certain economic activities or certain economic areas, where such aid does not adversely affect trading conditions is covered under Article 87(3)(c). As can be seen in Figure 4.1, the distribution of 87(3)a status covers rural areas that are relatively depopulated, but the 87(3)c status clearly focuses on post-industrial urban areas of the Midlands, north east and the north west of England and Scotland.

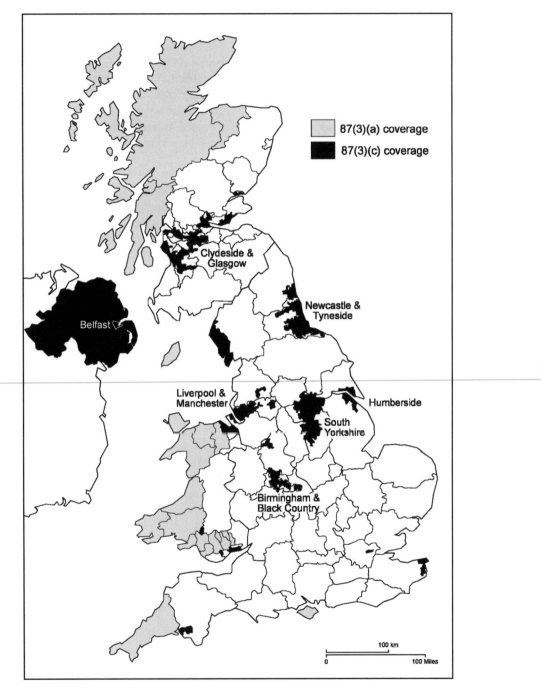

Figure 4.1   Map of Assisted Areas in Great Britain showing the areas approved by the European Union to receive regional aid from the government from 2007 until 2013. The designation of these areas indicates which regions have abnormally low standards of living.

Other funding sources

Three other funding sources are worth briefly considering: private investment (including Private Finance Initiatives), tax breaks and the National Lottery.

As discussed above, Area Based Initiatives use public funding to lever in private investment. However, private developers are not primarily interested in funding urban regeneration projects to alleviate poverty or create jobs for the local population. As private companies they are driven by the need to make a profit, either for their owners or, if publicly floated on the stock exchange, for their shareholders. A project must be economically feasible to attract private sector investment.

Many urban regeneration projects tend to be characterised by higher levels of risk than regular development projects. For example, the risk of developing a greenfield site in a desirable area of the south east is far lower than that of undertaking a **brownfield** development in a run-down urban location in a regionally disadvantaged area such as the north east. The potential costs and investment of time required to remediate brownfield sites can be high, and there is no guarantee that there will be sufficient demand for the premises when they are finished. Furthermore, the scale of many urban regeneration projects means that they can take years to complete, making them vulnerable to swings in the highly cyclic construction and property sectors. As Adair et al. (2003: 1075) state, 'the rules of the market inevitably encourage developers to go to the least difficult sites'.

The Urban Task Force (1999) report recognised that large-scale regeneration could not be financed by short-term debt or the public sector alone, and suggested the use of limited-life partnerships between the public sector and private developers in order to share the risks and long-term **capital** rewards. An example of a limited-life partnership is the Birmingham Alliance, set up by developers Hammerson, Henderson and Land Securities, to redevelop the city centre, including the Bullring. The need to mitigate risk also drives the preference of developers to take on large land packages, as regenerating a more comprehensive area gives them more control over the process, creates economies of scale and allows them to benefit from the multiplier effect, whereby a successful development can increase the potential profitability of surrounding sites.

The higher risks associated with urban regeneration projects have tended to deter institutional investors. Although they constitute a major source of private investment money in the UK, they are highly risk averse. In order to offset this level of risk, companies funding regeneration cut deals that will allow them to achieve above-average returns in exchange for financing projects. Where financial backing is required for a large project, commonly a development company will be set up and the financial backers will transfer funds by purchasing equity in the company. Typically, the development company will be responsible for insuring against the risk of non-completion. A limited number of institutional funds have been set up specifically for regeneration investments, such as the Igloo Fund within Morley Fund Management, although they tend to be more involved with secondary development and leasing than actual construction.

Private Finance Initiative (PFI) is a mechanism to attract private investment in capital and infrastructural developments. Introduced in 1992, PFI involves forming a partnership between a public and private organisation in order to fund a new asset that would normally be provided by the state. PFI generally involves the private sector designing, building and/or operating an asset, and the public supposedly gain from more efficient resource use and management. This has included schools, roads, hospitals, prisons and social housing schemes. PFI not only secures private funds to deliver public services, but shifts risk from the public sector into the private sector. Overall control lies with the private sector, with the role of the public agencies being to secure social benefits. Improvements to infrastructure can attract further investment into run-down areas, and open up new sites for development. PFI represents a way to use private sector finance to deliver public services.

PFI can be freestanding, whereby the private sector constructs and runs an asset, recouping its costs by selling the service back to the public sector. It can also involve the private sector simply gaining revenue through the running of the asset, or can take the form of a joint venture, whereby the private and public sectors contribute money to either a capital build or service delivery programme. In cases of exceptional cost, PFI **bonds** can be issued to raise funds. PFI also forges longer-term involvement of the private sector in development projects, which can increase the emphasis placed upon environmental considerations, such as the costs of running buildings over their entire life rather than simply the short-term costs of construction (a point that is returned to in the next chapter). PFI has been widely criticised in some quarters, as public assets (such as, for example, buildings and land associated with hospitals) are sold to private companies, who then lease them back to the public sector for a fee. In some cases these arrangements have seemed highly favourable to private business while compromising the public asset base; most notably it has been notoriously hard to pin down exactly how much risk the private sector actually bears in reality.

A slightly different form of funding involves the use of tax-based incentives to encourage development in areas where it is needed most. The Urban White Paper (DETR, 2000) highlighted the potential to use the tax system to influence property markets, both in terms of development, investment, ownership, letting and dwelling. Tax-based incentives include tax relief measures to make inner-city projects more appealing, such as capital allowances, capital gains tax relief on eventual sale, limited rate-free periods and **remediation** relief. Increased capital allowances and capital gains tax relief in particular are intended to free up more cash to fund further land and property projects. Reduced stamp duty and remediation relief make it more attractive for developers to purchase old industrial property and, where necessary, to clean it up for use, and reduced business rates provide an incentive to attract tenants.

Finally, the National Lottery allocates 28p out of every pound spent to 'good causes'. These cover a range of areas, from the arts and sport to heritage and communities. The funding bodies are split into 14 organisations, including those with a specific focus (for example, the Arts Councils for England, Wales and Northern Ireland), and more general funds (for example, the Millennium Commission, the Big Lottery Fund and the Heritage Lottery Fund). Lottery funding tends to be used to help build flagship regeneration projects such as the Lowry Centre in Salford (arts and heritage complex, Figure 4.2), Millennium Tower in Portsmouth (visitor attraction) and the Gateshead Millennium Bridge, which was the first opening bridge to be built across the Tyne for 100 years.

Evaluating Area Based Initiatives

ABIs have constituted the main policy approach to urban regeneration over the last 30 years. In the six competitive funding rounds between 1995 and 2001, 1,027 bids were approved under the SRB, equating to over £5.7bn in SRB support and £26bn of total

Figure 4.2   The Lowry Centre, Salford, which opened in 2000, is an example of a flagship arts capital project which was heavily supported with National Lottery funds distributed through three different agencies: the Arts Council of England, the Millennium Commission and the Heritage Lottery Fund.

expenditure across England (Rhodes et al., 2003). In examining the impact of delivery mechanisms over a two-decade period (1981–2000), Rhodes et al. (2005) estimate that the public sector spend on regeneration policy measures has been close to £10bn, which in turn attracted a £38bn spend by the private sector and other agencies. The estimated outputs of this investment have been nearly 18,000 hectares of reclaimed land, 22 million square metres of floor space, 350,000 net jobs and close to 195,000 new housing units.

On one level, the impact of these schemes is obvious, as they have transformed the actual physical appearance of target areas. In economic terms the impacts are harder to measure. Most cities have numerous projects occurring simultaneously, so any positive economic benefits will be the result of multiple factors. Further, it is impossible to predict how a city's economy may have developed in the absence of a particular regeneration scheme (Noon et al., 2000). Rhodes et al. (2005) identify three analytical problems in assessing ABIs. First, while the idea of focusing investment and resources on an area makes intuitive sense, the actual processes that are supposed to drive change are poorly understood. For example, it is assumed that the process of trickle-down will occur, but in practice there is a danger that ABIs merely displace investment from elsewhere in the city. Secondly, the tools used to evaluate ABIs have been poorly developed, partly due to the tendency of these initiatives to address a range of problems, from the provision of new infrastructure to job creation and crime prevention. Thirdly, and related to the diversity of problems addressed, there is a lack of data available to fully assess key goals against their outcomes. These problems have led some to argue that ABI regeneration has been largely superficial and has failed to address the underlying socio-economic problems of cities, merely displacing them. The next section turns to some real-world examples in order to see how economic regeneration works in practice.

Key points

- ABIs constitute the dominant approach to regeneration, concentrating funding in specific areas in order to create trickle-down effects in surrounding areas.
- Regeneration initiatives have shifted in emphasis from being purely property-led to seeking the integration of social and environmental factors, often through partnerships between different public and private organisations.
- Funding is competitive and primarily administered at the regional level through the RDAs.
- A range of other funding sources exist, including the European Union, Private Finance Initiatives and Lottery funding.
- Developers are risk averse and tend to prefer large-scale development opportunities.

Regenerating cities in practice

It is possible to identify Area Based Initiatives in most major cities in the UK. Projects range from the refurbishment of a single derelict building to the regeneration of entire areas of the city. It is worth considering a few examples to demonstrate the different types of project that have been used to jump-start economic growth in previously run-down urban areas. This section will also explore how cities balance the need to compete to attract generally similar kinds of industry, people and functions, while also incorporating their own unique identities and legacies into regeneration schemes.

Individual regeneration projects: Manchester Printworks

The regeneration of Manchester's city centre was given a unique kick-start in 1996 by a 3,300lb IRA bomb, which destroyed much of the central shopping area. The devastation prompted a £750m investment spree in the city, replacing a somewhat out-moded retail core with a host of fashionable developments – today a Harvey Nichols retail outlet stands on the site of the IRA bomb blast.

The Manchester Printworks building had been the headquarters of a number of national newspapers, occupying a prime city centre location on Corporation Street, but had stood derelict after the demise of Robert Maxwell's newspaper empire in the early 1990s. Refurbished at a cost of £150m, the Printworks was marketed as Europe's first urban

Figure 4.3   Manchester Printworks is a mix of generic restaurant and entertainment chains brought together, unusually, in a unique building. This combination has lent some character to a development which might otherwise be rather bland.

leisure and entertainment complex, covering approximately 32,500m^2 of floor space and housing 35 themed bars, 14 food outlets, a health complex, and the second largest Imax screen in the country (Figure 4.3).

The cost of refurbishing old industrial buildings is generally high, and this is the main reason that they tend to remain derelict. The internal layout of floors and walls had to be completely redesigned in order to accommodate retail and leisure developments, although the original layout of the Printworks makes it a unique space for this kind of development. Basic infrastructure, such as electricity wiring and water supply, needed to be re-installed.

Despite these costs, it is desirable to retain these buildings as elements of industrial heritage, as local communities associate them with a city's former pride. Analogues of the Printworks can be found in almost every city, such as Met Quarter in Liverpool and the Mailbox in Birmingham. They are also valuable as unique buildings that can be retained among otherwise generic developments, enhancing the identity of a city and its ability to attract potential visitors and businesses. The Printworks development is indicative of the importance of leisure uses to urban regeneration. Throughout the 1990s Manchester City Council vigorously embraced this approach to regeneration (Robson, 2002), promoting a very positive vision of the city based around leisure, from the Commonwealth games and its internationally renowned football teams to the 'Madchester' music scene that emerged from the Hacienda nightclub and bands such as the Happy Mondays. In 2001 it was estimated that around seven restaurants and bars were opening each week in Manchester, not an inconsiderable amount in a city of half a million residents. The opening of Manchester Printworks in 2000 marked the final phase in the rebuilding of Manchester's retail centre after the bombing.

Major regeneration areas: Laganside in Belfast

Waterfront developments have become a dominant theme of urban regeneration in the UK (see Chapter 6). As industrial ports have shut down, waterfront areas have become a focus of dereliction, but offer clear design and marketing opportunities associated with waterside locations. One of the largest regeneration projects in Northern Ireland focused on the Lagan River that runs through Belfast.

Belfast was little more than a village before the industrial revolution, which transformed it into one of the world's major ports. Over the course of the 1970s and 1980s the area around the Lagan had deteriorated due to the decline of the shipyards, and the area suffered from all the classic symptoms of urban decline. Socially, high levels of unemployment led to depopulation and deprivation, while physically the area was dominated by derelict buildings and low-quality housing. Even the Lagan itself had become an environmental liability, being heavily polluted and exposing foul-smelling sandbanks at low tide.

As the map in Figure 4.4 shows, the Lagan cuts straight through the centre of Belfast, and the scale of the blighted area (140 hectares) associated with its decline acted to severely retard the redevelopment of the inner city. While the economic potential of such a central area is obvious, the task of regenerating such an extensive and environmentally degraded area required large-scale action. With this in mind, the city drew up the Laganside Concept Plan in 1987 to outline options for rehabilitating the River Lagan and developing the adjacent areas. In 1989 the Laganside Urban Development Corporation was formed in order to begin the social and economic regeneration of the Laganside area. Their remit was extended to include the Cathedral Quarter nearer the city centre, bringing the total Laganside area to 200 hectares.

The Laganside Corporation used government money to encourage private investment in business and leisure, identifying key developments and working with private partners to deliver them. So, for example, the Laganside Corporation contributed £6m of funding to the construction of the Belfast Hilton Hotel and an undisclosed amount to the development of the McCausland Hotel. Flagship cultural projects such as the Belfast Odyssey Millennium project and associated facilities at Belfast Harbour were funded using approximately £90m of government money. The Odyssey was tagged as Northern Ireland's 'Millennium Project' and almost half of the funding came from the Millennium Commission, 11% from the Laganside Corporation, 18.5% from the Department of Culture, Arts and Leisure, 18.5% from the Sheridan Group and 3.3% from the Sports Council. In addition to the Odyssey project, a range of cultural venues have been developed, including the Waterfront Hall and Lanyon Place. The redeveloped areas have attempted to mix land use. For example, at May's Meadows, 140 apartments have been built, including 48 housing units provided by a housing association for tenants who qualify for social housing, in the same development as a bar, restaurant complex, call centre, the Hilton Hotel and the local headquarters of PricewaterhouseCoopers.

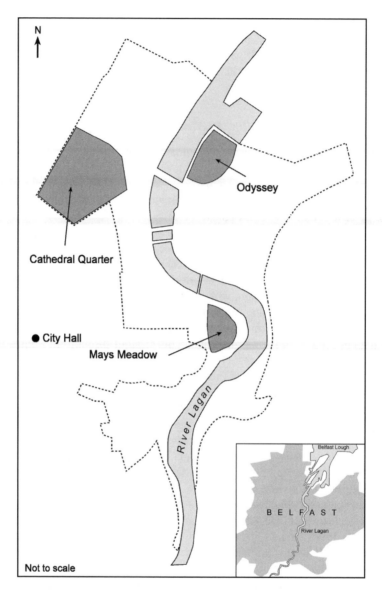

Figure 4.4   Laganside has been one of the longest-running urban regeneration projects in the UK. The map below shows the River Lagan, which cuts through the centre of the city, and the large areas of regeneration that have taken place around it. The revitalisation of Belfast has to a large extent depended upon the regeneration of this central river corridor.

Laganside has been one of the longest running regeneration projects in the UK; between its inception in 1989 and completion in 2007 it has secured:

- £1bn of investment;
- 14,700 jobs;

- over 213,000m^2 of office space;
- over 83,000m^2 competed retail/leisure space; and
- over 700 completed housing units.

The Northern Ireland Department for Social Development has recently taken over control of the Laganside area, as they consider that the development corporation has achieved its regeneration remit.

Developer-driven regeneration: Gunwharf, Portsmouth

In some cases major redevelopment projects are taken on by a single developer and it is worth examining an example of where this has taken place to understand the economic trade-offs made between local authorities and developers in practice. The Gunwharf development in Portsmouth was a major redevelopment of part of the derelict port, undertaken by Berkeley developers in the late 1990s (Cook, 2004). Established in the seventeenth century as the Naval Ordnance Department, Gunwharf employed over a fifth of Portsmouth's workforce in 1931. Successive reductions in the size of the British Navy, coupled with a concentration of destructive power in fewer armaments, led to the abandoning of the site in the mid-1980s. Portsmouth suffers from a series of problems, including depopulation, low levels of education and household incomes below the national average. In the early 1990s, Portsmouth City Council identified the port as an area ripe for property-led regeneration, comprising the usual mix of retail, housing and leisure.

In 1995 the City Council and Ministry of Defence announced that the Gunwharf site was to be put out to tender for private development. The Berkeley Group were awarded the contract as they were willing to part finance the proposed Millennium Tower, a landmark building to act as a focus for the port and mark the year 2000. The other developers refused to agree to part finance the tower, considering it to be uneconomic. The development was not guaranteed to be financially successful, however, and Berkeley negotiated a series of modifications to the initial brief with the City Council. They demanded to be allowed to build a designer shopping outlet in order to provide some form of economic 'insurance' in case the residential element of the development encountered problems. The City Council had been against any major retail outlet, fearing that it would take trade away from the other shopping centres in Portsmouth, but eventually agreed as they needed Berkeley to finance the broader regeneration of Gunwharf. Berkeley also claimed that the more upmarket outlet would be complementary to the city's other retail centres, appealing to local rivalry by claiming that it would elevate Portsmouth's retail GDP above that of their close neighbour and rival Southampton.

The 165-metre Millennium Tower was one of only 18 landmark projects to receive Lottery funding and initially Berkeley were going to top up the £30m of Lottery money with a further £9m. Controversially, they reduced their contribution to £3m in 1998, as they believed that the tower would not attract as many visitors as the Council claimed. While the intricacies of how the Portsmouth Millennium Tower was financed are too detailed to go into here, the point is that the developer was able to dictate terms to the city. For similar reasons, Berkeley managed to refuse to include any social housing in the first phase of development, flying in the face of city policy. Private developers thus play a massive role in regeneration projects and can wield significant power within Public Private Partnerships (PPP), especially when a project is driven by a single developer.

Flagship projects and branding

From the account given so far, it can be concluded that the external forces acting upon British cities are relatively generic. Their decline was due to the common problem of the waning competitiveness of UK manufacturing industry and their salvation is to be found in reinvention through regeneration. The broad goals of regeneration are thus shared. Cities need to enhance their appearance and liveability, improve their workforce and revamp their physical infrastructure. Within this general context, cities have begun to realise that in order to compete effectively they need to establish their own identity. In order to 'stand out from the crowd' and attract a share of the global knowledge economy it is necessary to establish a positive identity, by either building upon existing (perhaps neglected) heritage or by establishing new selling points.

Urban regeneration is critical to this process of generating an identity, and a range of strategies exist. For example, some cities have emphasised cultural attractions and industries, while others have sought to establish a reputation as world-class sporting venues. Some cities have focused upon attracting tourists, while others have concentrated on becoming business or shopping destinations. Increasingly, companies and planners seek to 'brand' their developments as if they were a product, to create a favourable identity that will enable them to be sold to potential residents or businesses. The idea of selling cities is not new (Ward, 1998), and is closely linked to the necessity of competing to attract people and business. Branding is now standard practice for individual developments (for example, the Park Central example in Chapter 3 involved extensive rebranding to change perceptions of an old council estate for marketing purposes). Increasingly, the marketing logic of branding regeneration is being extended to strategic, citywide partnerships. For example, 'Liverpool 1' (Figure 4.5) was launched as the brand for the £900m retail-led regeneration of Liverpool's city centre, unveiled in 2005. Liverpool 1 is an enormous scheme, covering six districts of the city, involving 30 individually designed buildings, 150,000m^2 of retail space, a 14-screen cinema, 21,000m^2 of restaurants, more than 450 new apartments, two hotels, offices, a revitalised 2-hectare park and a new public transport interchange. The PR friendly logo, selected after months of intensive marketing research, and catchphrase 'love the city', is used to market the development to consumers and retailers through a variety of channels.

ABIs often seek to incorporate so-called 'flagship developments' that establish an identity or brand. Often the facilities build upon local heritage, such as the BALTIC art gallery in Gateshead (these forms of cultural-led regeneration are addressed in more detail in Chapter 6). At other times, flagship buildings are parachuted into the city, like the Cardiff Bay Visitor Centre ('The Tube'), an award-winning building by Will Alsop that is designed to look like a futuristic telescope looking out across the bay. The Laganside development was littered with key flagship developments, such as the Odyssey Millennium project, while developments such as Manchester Printworks constitute standalone flagship buildings in their own right. In each case, these developments provide a symbolic and physical focus for an area's regeneration, and are often accompanied by a suite of less ambitious commercial developments around them. As with all ABI-led regeneration, the rationale for

LIVERPOOL
ONE

RULE 3
Love the City

Figure 4.5   The Liverpool One brand logo used to impart a clear identity to the regeneration of Liverpool's central shopping areas. The logo is typical of the current trend to brand redevelopments like products in order to attract potential investors, shoppers and residents.

flagship projects is that a trickle-down effect will occur, whereby economic benefits will gradually spread to surrounding areas. Whether and how this process actually occurs in practice is highly contested, as the example of Portsmouth's Millennium Tower demonstrates. But despite the gloss and glitz of many regeneration projects, they have been criticised in some quarters for being superficial and failing to address underlying economic issues. The example of Glasgow offers an insight into some of the issues surrounding economic regeneration.

Glasgow – critiquing economic regeneration

Glasgow was at one point the second city in the British Empire, acting as a hub for the massive concentration of shipbuilding and heavy industrial activities in the Clydeside region.

The decline of the Empire, coupled with deindustrialisation in the mid-twentieth century, saw massive job losses and the politicisation of a highly unionised left-wing workforce. The population halved from its peak at 1.3 million in 1930, and by the 1970s Glasgow was officially recognised as the most deprived locality in Britain, associated with a rather undesirable public image of crime, slums and heavy drinking. In 1981 the Glasgow District Council established an Economic Development and Employment Committee charged with reversing the city's economic fortunes. The committee quickly focused on rebranding the city with the 'Glasgow's miles better' campaign in 1983, and established a series of arts and cultural projects. At around the same time they began to pursue private sector participation in the physical regeneration of the city centre, and set up Glasgow Action, a business-led quango whose goal was to make the city a more attractive place to work, live and play (MacLeod, 2002).

A series of high-profile flagship regeneration projects were subsequently undertaken in key areas, including designer retail developments at Princes Square and the Italian Centre, café culture and gentrification within the Merchant City, and the transformation of Buchanan Street into a focal point for culture, shopping and leisure. On its own terms the regeneration of Glasgow has undoubtedly been a success. Named European City of Culture in 1990 and British City of Architecture in 1999, Glasgow now attracts thousands of tourists and shoppers each year and is a major conference location.

Despite the glitzy transformation of large parts of the city centre, doubts remain over the degree to which the city's inhabitants have benefited. For example, 100,000 people in the city remain dependent on state benefits (Scottish Executive, 2007), questioning the degree to which the wealth has trickled down from the regenerated areas to the rest of the population. As Marxist geographer David Harvey has noted, flagship regeneration projects like convention centres and art galleries are enjoyed more by wealthy visitors than the locally disadvantaged populations who live there. Related to this, the regenerating city can be seen as a patchwork of spaces that have become detached from one another. ABIs have completely renovated and transformed parts of the city, while neighbouring areas remain physically derelict and socio-economically deprived, questioning the logic of trickle-down.

Gordon MacLeod (2002) notes that the splintering of the spaces in the city is often accompanied by a social splintering, as certain types of people are excluded from the newly regenerated spaces. Again drawing on the Buchanan Street area of Glasgow, he describes how hostels in the area for the city's homeless have been cleared away as part of the refashioning of the space. Beggars, the homeless and street artists have no place in the newly imagined spaces of urban regeneration, which emphasise consumption through shopping, eating out and leisure. Those elements of society who lack the money to be able to partake in these activities are actively excluded by private security forces, the police and CCTV. MacLeod argues that the most dangerous feature of this process is the normalisation of this discourse through the popular press. Rather than arguing that the socio-economically marginal deserve help, the press tends to suggest that these unwanted elements of society are damaging the city's image and need to be removed or hidden from view. The economic success stories that are proudly proclaimed on regeneration project websites hide the story of who gains and who loses.

These problems are undoubtedly a side-effect of the neoliberal logic that champions competition and markets to deliver change. It seems almost inevitable that welfare and wealth redistribution agendas will become marginalised in favour of private commercial wealth generation. But despite the rhetoric of Public Private Partnerships, the local state has often borne the main financial risks in delivering flagship regeneration projects, while private industry has reaped the rewards. These issues are addressed more fully in the next chapter.

Despite these problems, there is little doubt that the ABI approach to urban regeneration has succeeded in reversing the economic decline of cities in the UK. Returning to Table 4.1 (see p. 53), the population trends for the UK's main regenerating cities are either stabilising or have become positive. As the case studies above show, urban cores have been revitalised. City-centre land is now attracting higher rents, business has been wooed back into the city and with it has come a skilled and more affluent workforce. But there are dangers in depending on attracting business to locate in a city, not least because they can just as easily leave if another city becomes more attractive. On top of this, in a scenario where most of the regenerating post-industrial cities of the UK are competing to attract similar types of industry and commercial use, there is a danger that there may not be enough to go around. The spectre of a national economic recession hangs over the successes of economic regeneration in UK cities. A second major strand of UK economic regeneration policy has thus attempted to generate organic economic growth within cities, by encouraging entrepreneurship.

Key points

- Public funding is used to subsidise early developments in large-scale regeneration projects, which then attract further investment.
- Flagship buildings are used to make powerful visual statements about regeneration projects that will put them on the map.
- Branding and image are increasingly central to regeneration partnerships and are used to market schemes to developers, business and the public.
- While hard to measure in quantifiable terms, urban regeneration has generally been judged an economic success, although social critiques of the neoliberal approach highlight the uneven distribution of benefits.

The entrepreneurial city

The knowledge economy and national policy

Economic policies to unleash the latent or internal potential of cities can be split into four related categories:

- improving the knowledge base;
- encouraging enterprise;
- education and training; and
- empowering local businesses.

Although these elements of economic policy work together, it is worth addressing each in turn as they frame current economic growth policy in the UK.

The UK government embraced the idea of the knowledge-based economy in the White Paper *Our Competitive Future: Building the Knowledge-driven Economy* (DTI, 1998). The paper argues argue that all businesses will have 'to marshal their knowledge and skills to satisfy customers, exploit market opportunities and meet society's aspirations for a better environment' (DTI, 1998: 6). Two ways in which current UK policy has attempted to harness high-value, knowledge-based industries include encouraging links between universities and industry, and cluster policy. The UK government announced a ten-year science and innovation investment strategy in 2004 that is designed to help the UK exploit the commercial opportunities offered by new technologies such as micro and nanotechnologies. The potential economic benefit to the UK is considerable, with the creation of high-value jobs and industries. Many cities are seeking to develop and attract these types of new technology through expanding the higher education sector and encouraging knowledge transfer between universities and high-tech industry. The international model for this approach to economic regeneration is well known: located on the outskirts of San Francisco, Stanford University played a vital role in educating the internet entrepreneurs who went on to make Silicon Valley one of the most economically successful regions on the planet.

One of the key policy responses to encourage this sort of economic development has been cluster policy. Based upon the work of Porter (1990), the DTI (2004: 6) defines clusters as 'concentrations of competing, collaborating and interdependent companies and institutions which are connected by a system of market and non-market links'. The idea underpinning clusters is that similar industries will locate in close proximity to one another in order to facilitate various linkages, for example, through the exchange of ideas, goods and workers. The DTI (now called the Department for Business, Enterprise and Regulatory Reform, DBERR) identifies three 'critical success factors', which clearly resonate with the wider tenets of the knowledge economy: the presence of functioning networks and partnerships; a strong innovation base with supporting R&D activities where appropriate; and the existence of a strong skills base. Connectivity between cities within regions is increasingly seen as one way in which to promote clustering, as it can make labour more mobile and boost the efficiency of competition by increasing the areas across which companies can operate. Much has been written about the tendency of high-tech firms to cluster and the regional development agencies (RDAs) have all followed the DBERR in adopting strong policies to encourage clustering. As major players in urban regeneration, the RDAs often apply the logic of clusters to developments in cities. This can take the form of earmarking new business parks as 'biotech-parks', or attempting to create high-tech corridors (essentially linear clusters) along key transport conduits. While clearly influential in policy, the logic of clusters has been criticised. Perhaps of key importance to urban economies is the question of whether it is possible to generate new clusters of industry, given the need for a range of 'soft' networks and the lifestyle demands of the creative classes. A second key question involves the types of industry that are being courted. While every city covets a biotech cluster, few divert funds to support existing but less 'trendy' clusters, associated, for example, with logistics (transport) and manufacturing industry.

Enterprise

Increasingly, policy-makers are interested in how businesses start up and grow. Small and medium-sized enterprises (SMEs) have been important drivers of economic growth. Over the last 20 years these businesses have created two-thirds of all new jobs, more than two-thirds of the innovation in the economy and have accounted for two-thirds of the differences in economic growth rates among industrialised nations (Walburn, 2005). Immigration has also had a startling impact on entrepreneurial activity. Almost a third of all high-tech firms started in Silicon Valley between 1995 and 1998 were run by Chinese- or Indian-born engineers. In 1998 these businesses had more than 58,000 employees and sales of close to $17bn (Saxenian, 1999). Jane Jacobs (1985) has argued that only cities have the potential to innovate, as they are the only places capable of substituting goods and services that they import for things that they produce themselves. While the argument is fairly complex, it involves the realisation that only cities can generate a critical mass of supply chains and consumer demand for products and services. It is surely no coincidence that London lies in the south east, which has been the UK's most economically prosperous region in the post-war period. This region has produced approximately 33% more significant innovations than other English regions over the same period. It is also one of only ten 'islands of innovation' in Europe, defined as accounting for 20% of the national research and design budget, having a strong presence of both research institutions and enterprises. The key needs of this sector were access to skilled workers, proximity to international airports and general standard of living factors such as quality of housing and schools (Simmie et al., 2002).

Three government papers frame enterprise and innovation policy and show a clear development in government thinking that seeks to link knowledge and enterprise as the basis for global competitiveness. The DTI's Science and Innovation White Paper (2000) set a framework for the government's role as the key investor in the science base and the facilitator of collaboration between universities and business. The White Paper on Enterprise, Skills and Innovation (DTI, 2001) linked the importance of science and innovation to regional (and national) economic growth, with the need to raise skills as a key issue. A number of initiatives were announced to invest in innovation and new technologies, including e-business and the need to foster an environment for enterprise. Building upon these themes, the most recent paper, *Competing in a Global Economy: The Innovation Challenge* (DTI, 2003a), situated the importance of knowledge and invention within the global context, stating that:

> The creativity and inventiveness of our people is our country's greatest asset and has always underpinned the UK's economic success. But in an increasingly global world, our ability to invent, design and manufacture the goods and services that people want is more vital to our future prosperity than ever. (DTI, 2003a: 5)

Creating the conditions for enterprise is notoriously hard, involving a suite of educational, financial and other support services and schemes. New facilities are often located within central regeneration schemes, such as the Moorgates Croft Business Centre in Rotherham, which is discussed in Chapter 5. Common barriers to entrepreneurship have been identified as:

- the involvement of too many agencies and institutions;
- changing names and remits of institutions and lack of policy coherence;
- discontinuity of funding; and
- problems of access to finance.

These problems are often a consequence of the fact that public programmes in the UK are time-limited, designed to start up new businesses or to demonstrate a commercial opportunity to a recalcitrant private sector. Programmes tend to collapse as soon as public funding ceases, damaging the credibility of the sector more generally. One emerging solution to these problems involves Community Finance Initiatives, which involves a form of charitable lending to enterprises that would be considered too high risk to be able to access traditional bank loans (Bryson and Buttle, 2005). While regeneration that focuses on bringing in investment from outside can be measured in terms of tonnes of concrete, jobs created or private funds, it is harder to measure the outputs of schemes designed to unleash local enterprise. This explains the current government's penchant for figures tracking numbers of business start-ups and the fortunes of SMEs, as the number of these businesses provides an indicator of the health of local enterprise.

Education and training

Early urban policy focused upon the need to create jobs and opportunity. For example, the Department of Employment White Papers *Employment: The Challenge for the Nation* (1985a), and *Lifting the Burden* (1985b) emphasised the importance of creating jobs and a skilled labour force capable of doing them. The papers set out a decidedly neoliberal agenda of enterprise and competitiveness for labour policy, which was subsequently backed up by the Employment Department Group White Paper *People, Jobs and Opportunity* (Department of Employment, 1992). In order to address the needs of the labour market, Training and Enterprise Councils (TECs) were set up in 1990 in order to channel public money into re-skilling and entrepreneurial programmes. The TECs were partnership-driven, but strongly influenced by the needs of private industry, advancing the neoliberal agenda of 'improved competitiveness, for individuals and businesses' (Hart and Johnston, 2000: 136).

These policies were criticised for prioritising private economic development to the exclusion of other important factors, such as health, environmental quality and the needs of community groups. When the Labour government took power in 1997, they retained the neoliberal approach to employment policy, but married it to a concern for including the groups in society who were excluded from mainstream economic activity. In 2001, the Learning and Skills Council (LSC) was set up in order to further the social inclusion agenda and help socio-economically disadvantaged groups who were relatively unskilled and dependent on benefits. The LSC is a non-departmental public body that replaced the former Further Education Funding Council and TECs. They had a budget for 2006–07 of £10.4bn to cover adult education outside the university sector. Resonating with the goals of the national strategies on innovation and competitiveness, the LSC aims to:

- raise participation and achievement by young people;
- increase adult demand for learning;
- raise skills levels for national competitiveness;
- improve the quality of education and training delivery;
- equalise opportunities through better access to learning; and
- improve the effectiveness and efficiency of the sector.

As might be expected, their remit is driven by the needs of industry. For example, they run the National Skills Academies that respond to employer needs to deliver the skills required by each major sector of the economy. One problem with adult education is the difficulty of capturing skills that have been (or need to be) learnt 'on the job'. The LSC deal with formal educational courses and awards, and often fail to capture the skills that exist in a workforce. As discussed above with reference to enterprise, formal efforts to support economic growth struggle to be responsive and comprehensive enough to meet the rapidly changing demands of businesses.

The European Union has also committed itself to a ten-year strategy of reform for Europe's product, capital and labour markets. The aim, agreed by European heads of state and government in Lisbon in 2000, is to create a Europe by 2010 that will be 'the most competitive and dynamic knowledge-based economy in the world, capable of sustainable economic growth with more and better jobs and greater social cohesion' (DTI, 2003b: vii). In response to the Lisbon Agenda, European regional funding is now clearly focused on stimulating growth and jobs at regional and local level. For instance, 75% of the Regional Competitiveness and Employment objective funding is earmarked for the Lisbon Agenda in the UK.

The government's commitment to aligning knowledge with economic growth finds expression in urban regeneration projects that incorporate new university buildings, halls of residence and libraries into development projects. Similarly, many cities have designated 'learning' or 'knowledge' quarters in an effort to encourage knowledge-intensive land uses in specific parts of the city.

Empowering local businesses

The ability to attract and nurture private businesses is key to economic regeneration, and under the rubric of 'partnership' the role of the private sector has grown in delivering urban regeneration. This chapter has already discussed a range of ways in which local business is involved in regeneration, such as PFI, or in terms of working with planners to deliver developments. Their level of involvement is also being increased at the local level though, as neoliberal policy initiatives seek to empower businesses to take ownership of their own business environment. One such scheme that has been imported from the United States is that of Business Improvement Districts (BIDs). BIDs are locally-based initiatives that allow groups of businesses and property owners to improve their urban environment. They generally involve a group of businesses agreeing to pay a voluntary tax in order to address key problems in the local urban environment or improve it in ways that go beyond the remit of the local state. This can involve improvements to buildings and streets in order to enhance the residential, commercial or retailing values of the area. It may also involve more general improvements to security like extra street lighting, maintenance of communal areas and facilities, and in some cases promotions and special events.

BIDs are judged according to strict criteria, whereby they must be proportionate to the benefits received, equitable and affordable, and generate significant business-sector responsibility (Ratcliffe et al., 1999). The key advantage of BIDs is that they devolve power to the local level, allowing a more responsive mechanism for addressing problems and seizing opportunities. They also generate higher levels of buy-in among stakeholders as everyone contributes and create a secure source of funding that prevents 'free-riding' (Mitchell, 1999). Critics of the scheme argue that BIDs transfer power over the public sphere into private hands, through the private financing of services that might traditionally be expected to be provided by the state. This can generate spatial inequalities, as some areas cannot support a BIDs scheme although the BIDs model has been adopted by 16 countries on four different continents (Ward, 2006).

Questioning the entrepreneurial knowledge city

Policy-makers and urban planners seem to equate future economic growth with an ideology of enterprise and are fixated on attracting high-tech and other knowledge-intensive industries (Armstrong, 2001: 524). Transforming post-industrial cities into flourishing centres of enterprise and knowledge is not, however, as straightforward as policy would perhaps lead us to believe. Not only is ICT catalysing the dispersion of footloose companies away from city locations, but cities without an industrial legacy often enjoy an advantage when attracting new business. It would be foolish to suggest that cities such as Sunderland and Liverpool can compete on the same terms as places like Cambridge and Reading, which enjoy a heritage of knowledge resources and a pleasant semi-rural environment in close proximity to London. Regenerating cities often aim for the second tier of the service sector, seeking to attract call centres and regional offices (with some exceptions). Here, being able to offer cheaper housing and lower costs of living for a workforce may score more highly than the quality of the local golf courses. This is a strategy that is serving Milton Keynes well, as it continues to be the fastest-growing city in England.

Of course, the danger with this strategy is that companies may decide to leave just as quickly as they came. The recent trend for outsourcing call centres to India is one such example of how changing divisions in the international labour force can affect regional economies in the UK. In response, cities have focused on releasing their innate potential for enterprise, concentrating on encouraging the types of knowledge-intensive business associated with the service sector. This emphasis can neglect the fact that a conurbation like Birmingham and the Black Country has a large number of light industrial and construction companies still operating in relatively central locations. This disjuncture often makes it hard to retain the existing economic base while simultaneously attracting service sector industry (although to a degree service sector industry can develop to 'serve' existing industry). Each is accompanied by very different workforces and requires very different forms of urban infrastructure.

On a related point, many commentators have noted that while the rhetoric of enterprise and training pervades job policy, the jobs created by urban regeneration are often menial in nature, or what have been termed 'McJobs'. For example, the leisure industry offers some of the lowest paid, least well trained and least secure employment, but often make up the bulk of jobs created by new regeneration schemes. Furthermore, these jobs are disproportionately taken up by the poorer, ethnic minority and female populations of the inner city. David Harvey (1989) has argued that rather than spending millions of pounds in public money on flagship buildings that have no real purpose, this money could be better diverted into training and education schemes for disadvantaged communities to enable them to obtain better jobs. These points are developed in the next chapter.

Key points

- Enterprise and innovation are seen as key drivers of economic growth in cities.
- This has led to an emphasis on education and training, in order to create skilled workforces that can work in the new economy.

(Continued)

- Policies seek to establish clusters of knowledge-intensive industries, to encourage entrepreneurial start-ups and to empower local businesses.
- Not all cities can compete to attract the most desirable industries.

Conclusions

This chapter has focused upon the key mechanisms of urban regeneration that have been used to revive the economic fortunes of cities in the UK. For the most part, this has involved regenerating old industrial manufacturing cities that have suffered from dereliction, depopulation, deskilling and deprivation to be attractive to the service sector industry. This task is not easy – as Richard Florida (2002) notes, the economically successful cities of today have to compete to attract a globally mobile 'creative class' of workers. Cities such as Glasgow and Manchester have undertaken major physical regeneration programmes, focusing on the wholesale redevelopment of specific areas of the city. ABIs have generally featured flagship regeneration projects designed to 'put the city on the map' as an attractive and culturally vibrant place, focusing on central areas that have been abandoned by manufacturing industry. Projects generally seek to maximise pre-existing assets like historic buildings and waterfronts, and draw upon a range of different funding sources to generate a mix of retail, leisure, business and residential uses, often organised around flagship developments to promote the trickle-down of wealth. In addition to the ABI approach that hopes to attract industry to the city, regeneration has focused on unleashing the economic potential of its own inhabitants. These policies have focused on improving the knowledge base, encouraging entrepreneurialism, education and training, and empowering local businesses.

On balance, it is hard to argue against the assertion that Britain's cities are in far better economic shape now than they were 20 years ago, although there are caveats to this conclusion. As with most evaluations, it is hard to know what would have happened to urban economies in the absence of regeneration policies, or given different policy goals. While the importance of a generally benign global economic environment over the last ten years cannot be overstated, it is worth noting that in the context of different urban policies in the late 1980s even an economic boom failed to lift the UK's cities out of their decline.

These successes are not unproblematic, however. The agenda for urban regeneration has been dominated by neoliberal policies that emphasise the role of the private sector in generating wealth at the expense of welfare and social support. This has undoubtedly created tensions between economic policies and wider social and environmental goals. The mono-logic of 'global economic forces' tends to marginalise certain groups in society who do not or cannot fit into the skilled worker/consumer blueprint for which today's city living is designed. Furthermore, questions have been raised over the extent to which all cities can play the same game. For example, the number of young professionals able to purchase city-centre apartments and the amount of consumer spending power to support new retail outlets and leisure facilities is not infinite. The rhetoric of place marketing and flagship regeneration schemes belies the increasingly generic appearance of many of the UK's cities. Many cities cannot realistically aspire to achieve the knowledge base and environmental quality of a city such as San Francisco, or even Cambridge, in the near future. As Deas and Giordano note (2002), there is no inevitability about economic regeneration. While Manchester has performed well from a very weak asset base, its neighbour Liverpool has received far more funding and yet not achieved anywhere near the same level of economic success. In the future, cities may well have to differentiate their regeneration more sharply from one another.

For a city to genuinely reinvent itself requires the integration of different policy areas. Social cohesion is vital in making a place safe and tolerant of difference; education and inclusion of the disadvantaged is needed to create skilled workforces; and improved environmental quality is necessary to increase quality of life and attractiveness to the creative classes. None of these things will happen if regeneration focuses solely on generating wealth for private companies and individuals. While the ultimate goal of urban regeneration is to rejuvenate city economies, the logic of private wealth accumulation needs to be reconciled with the need to include different social groups and improve the urban environment and infrastructure. Achieving this balance is the critical challenge facing urban regeneration in the UK and the next two chapters consider the role that social, environmental and cultural factors play in this process.

Further Reading

For those interested in the wider relations between cities and national economies, Jacobs (1985) provides a good place to start, having coined some of the most influential ideas concerning the relationship between cities and economic growth. Florida's (2002) work on the Creative Classes is required reading for anyone intending to understand contemporary urban policy, while the edited collections by Begg (2002) and Oatley (1998) provide an excellent insight into the competitive cities debate. In terms of ABI schemes, the review papers by Rhodes et al. (2003, 2005) provide excellent overviews of both the schemes and insights into their successes and failings. The field of economic regeneration in general draws much of its inspiration from the local economic development literature, and Plummer and Taylor (2003) provides a useful theoretical overview of local growth within the global economy.

Begg, I. (2002) *Urban Competitiveness: Policies for Dynamic Cities* (Policy Press, Bristol).
Florida, R. (2002) *The Rise of the Creative Class and How It's Transforming Work, Leisure, Community and Everyday Life* (Basic Books, New York).
Jacobs, J. (1985) *Cities and the Wealth of Nations* (Random House, Toronto).
Oatley N. (Ed) (1998) *Cities, economic competition and urban policy*, SAGE, London.
Plummer, P. and Taylor, M. (2003) 'Theory and praxis in economic geography: "enterprising" and local growth in a global economy', *Environment and Planning C: Government and Policy*, 21: 633–649.
Rhodes, J., Tyler, P. and Brennan, A. (2003) 'New developments in area-based initiatives in England: the experience of the single regeneration budget', *Urban Studies*, 40: 1399–1426.

5 Sustainability

Overview

Sustainability has become a central concept in all discussions of regeneration in the UK. This chapter explores the policy framework of sustainable regeneration, focusing on its social and environmental dimensions.

- *Sustainable development and regeneration*: explores what sustainability means and why it is important to urban regeneration.
- *The political framework*: outlines the key policies that frame sustainable regeneration, including the UK sustainability strategies and Planning Policy Statement 1.
- *Social sustainability*: discusses the Sustainable Communities Plan and the challenges of social regeneration, focusing on housing and gentrification.
- *Environmental sustainability*: examines the specific environmental dimensions of urban regeneration, including the brownfield agenda and the Egan Report on construction.
- *Masterplanning and sustainability*: explores the potential of masterplans to deliver more sustainable developments.

Sustainable development and regeneration

The term 'sustainable development' is notoriously ill-defined and requires some clarification before moving on to consider what it means in the realm of regeneration. The World Commission on Environment and Development (WCED) defined sustainable development as that which 'meets the needs of the present without compromising the ability of future generations to meet their own needs' (1987: 43). This commonly used definition is underpinned by the notion of **equity**, which means using resources fairly to meet the needs of

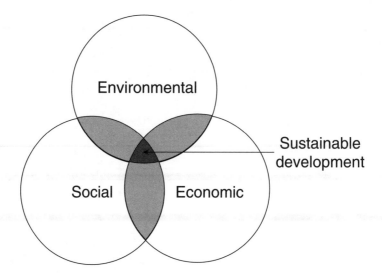

Figure 5.1   The three elements of sustainable development are often depicted as overlapping circles, with sustainability represented by the central area of overlap.

both current and future populations. Since the WCED report, considerable effort has gone into refining the concept of sustainable development. A common understanding of the term in policy is that it links economic, social and environmental concerns, represented as a set of overlapping circles, where sustainable development occurs in the central area of overlap (Figure 5.1). This tripartite definition has been likened to a three-legged stool, whereby if one of the legs is neglected, the stool will fall over. Another way in which the three elements of sustainability have been conceptualised is through the so-called 'triple bottom line', which extends the idea of an economic balance sheet to include social and environmental dimensions.

The goals of sustainable development are ideally suited to regeneration. Regeneration seeks to alleviate social, environmental and economic problems, through, for example, the re-use of derelict land and rejuvenation of impoverished housing stock (Couch and Denneman, 2000). Because Area Based Initiatives (ABIs) tend to address the needs of an area in a comprehensive manner over relatively long timescales, they offer the scope to address the three elements of sustainability in a joined-up way. Regeneration schemes often embrace the principles of sustainable development because they tend to attract public attention and enjoy a high political profile. Sustainability also emphasises participatory decision-making and partnership that can enhance the legitimacy of regeneration projects among local people. Roberts (2000) claims that the three pillars of regeneration are strategic vision, partnership and sustainability. The potential synergy between regeneration and sustainability is evident from the fact that strategic vision and partnership are also central to sustainability.

Since the Rio Earth Summit in 1992, UK policy has become framed by the goals of sustainable development. The principles of sustainability are filtered down from the national to the local levels across all areas of policy. This is certainly the case for urban planning and development. The mantra of sustainability lies at the core of the regeneration agenda and is loaded with expectations of being a magic bullet for delivering 'better' cities. Despite the wholesale adoption of sustainability in government policy, however, there is a great deal of debate and tension over what a sustainable development looks like in practice. The definitions given above are far from rigid. For example, the WCED version does not specify what constitutes a 'need', or over what timescale the future should be considered. In terms of the tripartite definition of sustainable development, different weightings can be given to each of the three components and there are obvious difficulties in comparing social or environmental gains and losses against more easily measured economic impacts. Different agents in the regeneration process have been able to use this uncertainty to place greater emphasis on one or other of the constituent components of 'sustainable' development, a problem highlighted by a number of authors.

While the complexities surrounding the various meanings of sustainable development have generated a literature in their own right (Mebratu, 1998), the result has been a certain amount of ambiguity over what sustainable regeneration looks like in practice. It is worth noting that 'sustainable development' does not always mean the same thing as 'sustainability'. Although the terms are often used interchangeably, sustainable development is a mainstream formal policy discourse, while sustainability keys into a wider set of ethical ideas that tend to emphasise social and environmental elements. This chapter explores the tensions that surround the process of making regeneration sustainable. It begins by outlining the key policies framing sustainable regeneration, before moving on to consider the social and environmental dimensions of regeneration.

The policy framework

The UK strategy

The UK government officially committed to the goals of sustainability at the United Nations Conference on Environment and Development Earth Summit held in 1992. In addition to signing a series of major treaties on climate change, biodiversity conservation and principles on forests, it signed up to the Rio Declaration and Local Agenda 21, which set out the procedures through which sustainability should be put into practice more generally. A key principle set out in these documents is that of *subsidiarity*, whereby international policy priorities are cascaded down through the national and regional tiers of government to the local level. This principle lies behind the Rio tagline of 'think global, act local'. Sustainable development thus rests upon a decentralist philosophy that asserts the need to plan at the local level. Sustainability advocates the inclusion of local stakeholder groups in decision-making in order to achieve action at the local level. It does not advocate a 'one-size fits all' solution, but recognises that decisions must be sensitive to the demands and opportunities of differing contexts. It is not about finding a perfect solution, but is about achieving a balance that brings the most benefit to the most people.

The holistic approach of sustainability fits neatly with agendas of inclusiveness, multi-agency partnerships and the shift from government to governance that have been pursued with great enthusiasm since the election of the New Labour government in 1997 (see Chapter 3). The government has published three strategies for sustainable development since the Rio Earth Summit:

- *Sustainable Development: the UK Strategy* (Department of the Environment, 1994);
- *A Better Quality of Life: Strategy for Sustainable Development for the UK* (DETR, 1999); and
- *Securing the Future: UK Government Sustainable Development Strategy* (ODPM, 2005c).

Four key elements run through these documents: social cohesion and inclusion, protection and enhancement of the natural environment, prudent use of natural resources, and sustainable economic growth. The most recent strategy emphasises that challenges such as tackling climate change require us to address these elements in a more integrated manner.

The general goals of the strategies for sustainable development were developed into a more specific vision for regeneration in the Urban White Paper *Our Towns and Cities: the Future*, which aimed to:

> ...bring together economic, social and environmental measures in a coherent approach to enable people and places to achieve their economic potential; bring social justice and equality of opportunity; and create places where people want to live and work. These issues are interdependent and cannot be looked at in isolation. For instance, there are close links between housing, health and education. That is why moving towards more mixed and sustainable communities is important to many of our plans for improving the quality of urban life. (DETR, 2000: 8)

As this quote indicates, the White Paper emphasised the holistic elements of sustainability, and the need to incorporate environmental considerations into all aspects of design.

Table 5.1 outlines the key dimensions of sustainable regeneration that were set out in the Urban White Paper (DETR, 2000). While Table 5.1 categorises the elements of sustainable regeneration as environmental, economic and social, it is clear that in practice each goal addresses more than one element of sustainability. For example, rejuvenating housing stock simultaneously improves the economic base of an area, benefits the inhabitants socially by providing a better quality of life, and enhances environmental quality. Similarly, each element of sustainability is addressed by more than one regeneration goal. For example, the 'prudent use of natural resources', which is emphasised in the UK strategy, is addressed by the re-use of derelict land (protecting the countryside), **mixed-development** (reduce car use and protect the atmosphere), energy efficient buildings (reduce consumption of energy) and higher density developments. The Urban White Paper makes very strong links between the idea of 'mixed' and 'sustainable' communities. This is a key element in relation to achieving socially sustainable regeneration. The White Paper also formalised the **brownfield** agenda, setting specific targets for 60% of new developments to be built on brownfield land by 2008.

These goals have been subsequently refined into a series of initiatives and approaches. The government's independent adviser on sustainable development issues, the Sustainable Development Commission, released a paper in 2003 on sustainable regeneration that identifies key initiatives that are used to deliver these goals. They identify factors such as energy efficiency in construction as critical in combating fuel poverty, which is a major issue facing poorer communities; local environmental action as a mechanism to create jobs and promote community re-investment; Green Transport Plans and Home Zones to improve public transport services, cycling and walking facilities and links to local services;

Table 5.1   Key elements of sustainable regeneration

Element of sustainability	Goal	Reason
Environmental	Re-use derelict land for high-density development	Protect countryside and decrease car use
	Improve environmental quality	Enhance quality of life and attract investment
	Use energy efficient buildings	Decrease ecological footprint of urban areas
Economic	Rejuvenate housing stock	Revitalise city centres
	Attract development and create jobs	Improve local economy
Social	Mixed-use developments (combination of retail, residential and business)	Decrease car use (live, work and play in same area)
	Mixed communities (in terms of age, ethnicity, family structure and income)	Increase social integration
	Inclusive decision-making	Respond to local needs and increase social capacity

and the need for re-investment as well as redevelopment, to avoid bulldozing housing stock that could be regenerated.

Quite quickly it has become clear that the goals of sustainable regeneration cannot be delivered by individual organisations or sectors. They cut across policy domains that have traditionally been seen as separate, such as transport and housing, and greatly exceed the organisational and financial capacity of the state. Sustainable regeneration requires planners, communities, developers, and architects to work towards similar goals, and national polices for sustainable regeneration have been translated into guidance for a range of organisations typically involved in the regeneration process. Planning is the hub around which regeneration partnerships are often formed, and before considering the social and environmental aspects of regeneration it is necessary to understand the role of the planning system in delivering sustainability.

Planning and sustainability

The planning system is the core mechanism to deliver sustainable regeneration. As Sue Owens (1994: 440) points out, 'planning and sustainability share two fundamental perspectives – the temporal and the spatial. Both are concerned with future impacts on and of particular localities.' The government is publishing a series of Planning Policy Statements (PPS) that guide local planning authorities on issues ranging from flood control and housing to green belts and geology. Sustainability forms the guiding 'vision' of UK planning policy (Davoudi, 2000), with PPS 1 *Delivering Sustainable Development* providing the overarching framework for all the planning policy statements (ODPM, 2005b) (although it is important to remember that the statement only applies to England – see HM Government (2005) for an overall framework).

PPS 1 identifies five areas in which planning can deliver the goals of the UK strategy for sustainable development:

- Making suitable land available for development in line with economic, social and environmental objectives to improve people's quality of life.
- Contributing to sustainable economic development.
- Protecting and enhancing the natural and historic environment, the quality and character of the countryside, and existing communities.
- Ensuring high-quality development through good inclusive design, and the efficient use of resources.
- Ensuring that development supports existing communities and contributes to the creation of safe, sustainable, liveable and mixed communities with good access to jobs and key services. (ODPM, 2005b: 9–10)

The biggest task facing the planning system is the need to facilitate economic development while also protecting the environment and people's rights and interests. This tension has been thrown into relief by the latest Barker Report, commissioned by the Treasury (Barker, 2006), which reviews the efficiency of the planning system in the UK. The report contains a number of suggestions to promote the ease with which developments may be approved, but seeks to reduce the level of environmental considerations that are taken into account and undermines the ability of local people to object to proposals. In polarising opinion between environmental and economic interests, the report highlights the challenges of balancing economic, environmental and social factors in practice.

PPS 1 also outlines the key mechanisms of *partnership* and *plan making* by which the goals of sustainability are to be delivered. They argue that this requires:

> ...a transparent, flexible, predictable, efficient and effective planning system that will produce the quality development needed to deliver sustainable development and secure sustainable communities. National policies and regional and local development plans (regional spatial strategies and local development frameworks) provide the framework for planning for sustainable development and for that development to be managed effectively. Plans should be drawn up with community involvement and present a shared vision and strategy of how the areas should develop to achieve more sustainable patterns of development. (ODPM, 2005b: 3)

This plan-led system is intended to provide the strategic capacity to balance the various demands of delivering sustainable development. This may involve, for example, ensuring that a suitable supply of land is made available for development while simultaneously conserving key environmental spaces within an area, in order to meet targets for both housing provision and greenspace provision. Regional and local development plans have sections addressing categories that cover each area of the PPS. In addition, further policy may be provided in Supplementary Planning Guidance (SPG). This may deal with specific policy concerns, such as nature conservation, or it may pertain to a local action plan for the regeneration of a specific area. Development plans and SPG constitute legally binding material considerations when the local planning authority or the development control committee decide whether to allow individual developments to proceed.

The plan-led system is intended to be transparent, efficient and inclusive. This means that the plan-making process is open to public review, allowing local stakeholders to contest policies and land-use designations, and that the planning authority must provide annual updates concerning the plan's implementation. In providing a level of predictability in terms of what kinds of development will be allowed where, this system is intended to provide more certainty for those delivering sustainable development. The system also encourages local communities and businesses to participate in drawing up strategies and plans for local areas. Inclusiveness is delivered primarily through Local Strategic Partnerships (LSPs), which bring together the public, private, business, community and voluntary sectors in order to address initiatives and services within local areas. While these partnerships echo Local Agenda 21 in terms of their collaborative basis, they are more important in terms of securing funds. Local Area Agreements (LAAs), which set out the priorities for a local area agreed between central government and key partners at the local level over a three-year period, are critical in accessing central government funding. Local Strategic Partnerships are often involved in setting goals for sustainability and improvements in quality of life. The role of communities in social regeneration is considered in more detail in the next section.

Key points

- Sustainable development means achieving a balance between social, economic and environmental factors. Sustainability frames UK policy at all levels.
- Urban regeneration is ideally suited to deliver sustainability, as it offers the potential to address the problems of an area in an holistic, long-term way.
- Sustainability cannot be delivered by the government alone. The planning framework for sustainability is therefore partnership-driven and plan-led.

Social sustainability

Sustainable communities

Within the overall policy framework for sustainable development, the government has placed increasing emphasis upon the idea of *sustainable communities*, as 'places where people want to live and work now and in the future' (ODPM, 2003: 56). The Deputy Prime Minister launched the Sustainable Communities Plan (henceforth referred to as the SCP) in 2003, setting out a long-term programme of action for delivering sustainable communities in both urban and rural areas. The transformation of the ODPM into the Department for *Communities* and Local Government (CLG) reflects this emphasis, adopting the vision of 'creating prosperous and cohesive communities, offering a safe, healthy and sustainable environment for all' as its over-arching mission statement.

The SCP translated the wider goals of sustainable urban development into actions for specific places. As with previous urban development agendas, the key element of the plan was to tackle housing quality and supply issues, but the SCP also prioritised the need to improve the quality of the **'public realm'** – the surrounding environment and community services that make an area more liveable. The SCP included a significant increase in resources for housing and major reforms of planning. Regeneration is fundamental to the strategy, providing a mechanism to build housing that people want, through the creation of affordable housing in the south east and improving housing in the disadvantaged areas of the Midlands and northern England. The integrated approach of regeneration allows issues of environmental quality, building design and economic development to be addressed in an **holistic** manner. The SCP identifies 12 dimensions of sustainable communities (see Box 5.1) that clearly follow the principles of social inclusiveness set out in the UK sustainability strategies and PPS 1.

Box 5.1   Principles of sustainable communities set out in the Sustainable Communities Plan

- A flourishing local economy to provide jobs and wealth.
- Strong leadership to respond positively to change.
- Effective engagement and participation by local people, groups and businesses, especially in the planning, design and long-term stewardship of their community, and an active voluntary and community sector.
- A safe and healthy local environment with well-designed public and green space.
- Sufficient size, scale and density, and the right layout to support basic amenities in the neighbourhood and minimise the use of resources (including land).
- Good public transport and other transport infrastructure both within the community and linking it to urban, rural and regional centres.
- Buildings – both individually and collectively – that can meet different needs over time and that minimise the use of resources.
- A well-integrated mix of decent homes of different types and tenures to support a range of household sizes, ages and incomes.
- Good quality local public services, including education and training opportunities, health care and community facilities, especially for leisure.
- A diverse, vibrant and creative local culture, encouraging pride in the community and cohesion within it.
- A 'sense of place'.
- The right links with the wider regional, national and international community.

Since 2003 the government has committed £22bn to infrastructure development projects as part of the SCP. Over five years £6bn has been invested in the Thames Gateway, while £1.2bn has been invested in the north of England and Midlands in areas which have large areas of low-quality housing. The main mechanism of housing renewal to improve run-down housing stock has been the Housing Market Renewal Pathfinders, and CLG estimates that they have refurbished over 13,000 homes. The SCP has attracted criticism on the grounds that it is no different from the urban development corporations, focusing on physical infrastructure with little consideration of sustainability in any broader sense. This criticism has been fuelled by the first Barker Report (Barker, 2004), which reviewed housing supply in the UK. The report recommended that more housing land be allocated in areas of high demand, leading to the majority of the budget being directed away from deprived areas in the north towards the over-heated south east, particularly through the Thames Gateway. Critics have argued that because the report was driven by the Treasury, it was more concerned with the need to reduce house prices in the south east than with the regeneration of disadvantaged areas.

The government attempted to redress the overriding focus on housing renewal in the original SCP with the publication of a five-year plan for sustainable communities. The report recommended that more power over how areas are planned should be devolved to the local level, with Deputy Prime Minister John Prescott stating that 'people live in neighbourhoods, not just in houses' (ODPM, 2005d, 2). The second Barker report (Barker, 2006), however, represented a shift back to a Treasury-driven agenda which prioritises economic development. The new Planning Act, 2008, in line with Barker, streamlines the planning system, but at the expense of reducing local people's ability to contest major planning decisions.

Turning from these macro-economic concerns to social issues, it has also been noted that while the SCP is clearly based upon the broader policy principles of sustainable development, it represents a very specific view of the role of communities in sustainable regeneration. Mike Raco (2005) identifies a dual discourse within the SCP. First, he argues that sustainable citizenship is equated to lessened dependence on the state. Communities are essentially being encouraged to engage in forms of self-governing and self-help with an emphasis upon training, entrepreneurialism and community stewardship. Secondly, sustainable citizenship is demonstrated economically through ownership of property. As the SCP states, 'owning a home gives people a bigger stake in their community, as well as promoting self-reliance' (ODPM, 2003: 37). This emphasis upon self-help and independence has more in common with **neoliberal** policies that emphasise the restriction of aid to those who demonstrate the capacity to help themselves than the principles of equity upon which sustainable development is based.

A further issue that Raco identifies within the sustainable communities framework involves how the problem of housing has been framed. In the south east the problem involves the inability of so-called 'key workers' (school teachers, nurses, firemen, etc.) to afford housing. In the Midlands and northern England the problem involves economically self-sufficient households fleeing inner-city areas for suburban or rural locations. While poor households have suffered from problems of unaffordable housing and inner-city blight for a number of years, in each case the driving factor is the middle classes. There is a perceived need to use regeneration to attract middle-class households back into these areas in order to improve them. Both of these issues raise the question of *who* is included and excluded from the idea of the sustainable community. This issue is explored in the next case study.

Figure 5.2   Attractive waterfront flats at Salford Quays, which sit opposite the Lowry Centre (Figure 4.2). This enclave of wealthy professionals sits close to areas of relative deprivation, raising questions about its social sustainability.

Case study: Salford Quays

Salford is located 4 km to the west of Manchester. The city's docks, built in 1894, serviced the Manchester Ship Canal, employing 3,000 people at its height. The industry gradually declined from the post-war period until 1982 when the docks were closed down. The derelict docks covered a substantial area (150 hectares of land and 75 hectares of water) and their closure compounded the problems of what was already a depressed area of Salford, typified by high levels of unemployment. Salford City Council purchased the land in 1983, realising that it represented an opportunity to transform a major eyesore into a focal point for development in Salford. They established a development plan for the area, which they branded 'Salford Quays'. The plan aimed to place equal emphasis upon work, leisure and residential land uses, as well as to provide infrastructure such as roads and key services. Emphasis was placed on the design quality of infrastructure, and 20 years, on Salford Quays is widely regarded as a model of successful regeneration, turning a large derelict area into a thriving mixed-use development (Figure 5.2). There are over 150 businesses in the Quays, 2,000 dwellings and 18,500m^2 of office space. Proposed residential blocks are becoming higher, indicating the growing confidence of developers in the appeal of the area.

Recent work by Raco and Henderson (2005) has highlighted a series of tensions in terms of how successful the development has been in creating a sustainable community. These can be categorised as internal, external and strategic problems. Internally, there has been a lack of attention paid to the needs of the residential community in Salford Quays. Commercial and residential development was prioritised at the expense of developing community services and spaces. This is perhaps understandable, as private finance provides the life-blood of regeneration projects, but has resulted in a situation where the projected quality of services available to the community has been compromised. The lack of clear strategic thinking in terms of what the resultant community might look like and how it would function has resulted in a fairly one-dimensional population of young professionals. The lack of focus on local public services, including education and training, health care and community leisure facilities, means that, like many urban regeneration projects, there is little provision for other types of potential resident, such as young families. Services like public transport need to be installed at the beginning of a development. If they are neglected or added as an afterthought, then the residents will have already made other arrangements, in this case to drive. Similarly, a young family will not move in to a development and 'wait' a number of years for a nursery to be built.

Externally, the Salford Quays community is relatively isolated from the surrounding areas. At the most basic level, the regenerated area now represents an island of owner-occupied housing within a predominantly publicly-owned housing area of Salford. On the one hand, advocates of trickle-down theory would argue that the area will serve to raise the expectations of the surrounding less advantaged population. In practice, however, the residents in the surrounding area feel very little attachment to the development as a whole, as the people who have moved into Salford Quays tend to be wealthier and mix very little with the pre-existing communities surrounding the development. This absence of trickle-down benefits to local areas is a widely noted phenomenon (Healey, 1997). The division can often be starkly visible, as new developments directly adjoin run-down council estates. A noted feature of riverside regeneration is the tendency to create 'canyons' of wealthy regenerated areas, bounded immediately by derelict land and deprived areas. These tensions are expressed most acutely in terms of commercial and residential theft, and new residential schemes are characterised by high levels of security and surveillance. Indeed, many new developments boast of their secure credentials in terms of guards, CCTV and physical features such as fences and defensive planting in their marketing literature (see Atkinson and Helms, 2007 for a comprehensive review of security and urban regeneration). While various training and employment schemes were put in place in order to allow the redundant industrial workforce in the surrounding areas to access job opportunities in the nascent service sector of Salford Quays, these schemes have not proved sufficient to integrate the surrounding community.

Strategically, these challenges raise questions concerning the ability of regeneration projects to deliver sustainable communities. The rejuvenation of specific areas such as Salford Quays, set within larger areas of inner-city decline, attracts a new type of household into the area. This process represents an extreme form of **gentrification**, as more affluent people move into an area that has already been upgraded for them (for the classic analysis of gentrification in the city, see Smith, 1996; for another example, see Cameron, 2003). While this process may reinvigorate the local housing market, it can also have the effect of excluding the pre-existing poorer inhabitants from the ability to buy property in their own area. This would clearly be a problem given the emphasis of the Sustainable Communities Plan upon ownership of property as the bedrock of sustainable citizenship.

Indeed, the kinds of owner that are encouraged by the dominant development logic of one- and two-bed flats are limited to young professionals and buy-to-let investors targeting young professionals. In the terms of the Sustainable Communities Plan (SCP), this is not a well-integrated mix of homes of different types and tenures capable of supporting a range of household sizes, ages and incomes. Furthermore, young professionals are essentially migrant populations. They have to be highly mobile, both nationally and internationally, in order to chase career opportunities. This is almost antithetical to the qualities of 'a diverse, vibrant and creative local culture, encouraging pride in the community and cohesion within it' that the SCP preaches. Indeed, the government's focus on building high-density flats has exacerbated the chronic shortage of family homes in many parts of the UK.

These difficulties highlight the tension between physical regeneration as the transformation of the built environment and social regeneration as a community transformation. The first tends to involve attracting new people to an area and can exclude the pre-existing populations. The second is inclusive of the pre-existing population. This problem is partly related to the flagship status of regeneration projects, which means that their success is judged largely in terms of making derelict areas economically successful. Politicians and investors often demand 'early wins' – quick and highly visible transformations that distance regeneration projects from the pre-existing problems of an area and/or demonstrate profitability. As one planner in the Midlands stated while surveying the view of the city centre from the window of his eighteenth storey office, he can tell how well he is doing his job by the number of cranes visible. Physical transformation is often accompanied by symbolic transformation. The re-labelling of areas for the purposes of regeneration reflects a perceived need to dissociate an area's past from its future.

While not on the same scale as the *tabula rasa* (clean slate) approach to regeneration that characterised the wholesale slum clearance programmes of the 1950s and 1960s, the logic of 'build it and they will come' works economically by attracting a new population into an area. Whether this form of economic regeneration actually creates sustainable communities is highly contestable. The property-led approach of the SCP does not, in and of itself, deliver 'sustainable communities'. This begs the question of whose responsibility it is to ensure social integration. At present a massive burden of expectation is being placed on skills provision and training to bridge this gap, while the majority of the resources are being poured into physical regeneration.

Key points

- The Sustainable Communities Plan emphasises the role of regeneration in providing quality affordable housing stock, skills and amenities for disadvantaged populations.
- While sustainable communities are supposed to be economically viable, mixed and have a sense of ownership over the place in which they live, regeneration schemes tend to attract new populations to an area, such as young professionals, and there can be difficulties integrating new and pre-existing populations.
- There is a tension between the emphasis of regeneration on physical transformation and economic development and the social goals of creating sustainable communities.

Environmental sustainability

In its rhetoric at least, the SCP recognises that making places where 'people want to live' requires more than simply the redevelopment of housing stock and building new homes. The provision of a high-quality, liveable environment is also a prerequisite. The environmental agenda in the UK has two goals with respect to regeneration, both set out most clearly in the Urban White Paper (DETR, 2000). The first involves preserving the countryside by re-using derelict or contaminated urban land for development. The second is to ensure that new developments are environmentally friendly. This section outlines the key elements of each of these challenges.

The most basic characteristic of this agenda is that development is focused in cities in order to protect the countryside and rejuvenate urban areas. The separation of rural and urban affairs can be traced back to the Town and Country Planning Act of 1947, which founded the modern British planning system. While the history of this division falls outside the remit of this chapter, it is salient to note that the division still underpins national policy, with the government White Papers on sustainability being split into urban and rural papers. This separation between urban and rural policy has been criticised for contradicting the holistic ethos of sustainability. Cities are highly connected to the regions in which they are located through flows of people (e.g. commuters) and materials (e.g. water, food). Separating their planning functions means that cities lose 'control of the areas where their growth would naturally go' (Gracey, 1973: 77). This tension frames debates concerning urban and rural sustainability in the UK and undermines current calls to establish 'city-regions' as political units.

The primary development pressure upon land in the UK has been the need for new housing. Since the spread of suburbia in the inter-war period, housing has been seen as the primary cause of urban sprawl. Given rising demand, policy has focused on building a higher proportion of houses in urban areas, especially on brownfield sites. Many urban areas have a legacy of derelict and contaminated sites as manufacturing industry in the UK has declined. The idea of developing brownfield land is eminently sensible (Parliamentary Office of Science and Technology, 1998). Derelict land is often an eyesore that is associated with general urban decay – both social and physical – a condition termed 'blight'. People may be afraid to walk past or through such areas. It is also wasted space in terms of the potential rents that could be fetched due to its close proximity to urban centres and large populations. Filling in these 'gaps' in the urban landscape creates more compact cities that encourage lower car use. The Urban White Paper (DETR, 2000) proposed that 60% of new housing should be located on brownfield land by 2008, and prioritised the remediation and use of contaminated land for housebuilding and urban developments in general. The regeneration agenda in the UK has thus become synonymous with the development of brownfield land.

Figure 5.3 shows the percentage of total new residences built in rural and urban areas between 1985 and 2003. The graph clearly shows that the amount of housing development in urban areas has grown from under 50% in 1985 to over 65% in 2003, and that within the urban category it is vacant and derelict land that has accounted for this increase (Karadimitriou, 2005). Similarly, some of the tenets of sustainable regeneration have been adopted enthusiastically. For example, higher density developments have been popular as more units on a site allow developers to make more money, and the density of new

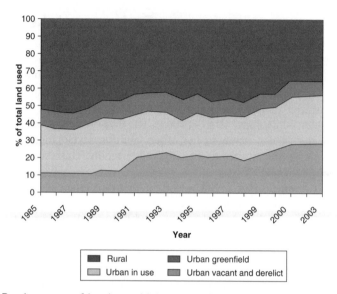

Figure 5.3   Previous use of land on which new residences are built, England 1985–2003. Note the increase in the proportion of previously used urban land to build houses. (After Karadimitrious, 2005)

dwellings built has increased substantially from 26 to 40 dwellings per hectare between 1996 and 2006.

At the same time, the Urban White Paper recognised that locating increasingly dense developments in cities could potentially exacerbate environmental, social and economic problems, unless attention was paid to the provision of quality public spaces. *Planning Policy Guidance 17: Planning for Open Space, Sport and Recreation* (ODPM, 2002b) identifies four emerging themes relating to open space policy:

- An increasing emphasis on the wide range of roles played by open space in enhancing the liveability of towns and cities.
- The importance of taking an holistic view of planning for green space in towns and cities, whether parks, river valleys, linear walkways or incidental open spaces.
- The importance of incorporating sustainable development principles which, for example, would increase the emphasis in favour of retaining and creating small accessible open spaces in towns and cities for use by local residents.
- A greater emphasis on the improved management of open space and greenspace networks.

This open space agenda is seen as critical in supporting the kind of urban renaissance that the Urban White Paper called for. As PPG 17 states:

local networks of high-quality and well managed and maintained open spaces, sports and recreational facilities help create urban environments that are attractive, clean and safe. Green spaces in urban areas perform vital functions as areas for nature conservation and biodiversity and by acting as 'green lungs' can assist in meeting objectives to improve air quality. (ODPM, 2002b)

In the same year, an ODPM report entitled *Living Places: Cleaner, Safer, Greener* (2002a) linked the open-space agenda to the problem of environmental exclusion, whereby poor local environmental quality is related to areas and households experiencing multiple deprivation, a significant factor in maintaining social exclusion and perpetuating cycles of deprivation. Within the current open-space agenda, the major emphasis is on the social benefits of the environment, such as inclusion, community cohesion, health and well-being.

The environmental challenge of brownfield development is thus twofold: to remediate the land for development, while ensuring that any valuable open space is either retained and enhanced in the proposed development, or replaced. Over time the problems of reclaiming brownfield sites have grown, due to the simple fact that the preferential development of uncontaminated sites has left an increasing proportion of contaminated sites relative to overall brownfield stock. The debate is complicated further by the broad range of land included under the umbrella term 'brownfield'. Brownfield land is land that has been previously developed, and PPG 3 (DETR 2000b) defines 'previously developed land' as that which 'is or was occupied by a permanent structure (excluding agricultural or forestry uses)'. As a number of authors have noted, this includes vacant, derelict and contaminated land and could range from an overgrown railway siding to an old colliery site contaminated by heavy metals (e.g. Alker et al., 2000). Despite attempts to distinguish between different types of brownfield site in policy, the terms are often used loosely and interchangeably in practice.

Some sites that are classified as brownfield are used informally by local residents for recreational activities, or have conservation value. The example of biodiversity nicely highlights this difficulty. Lord Rogers states in his introduction to the Urban White Paper that 'building more than 40% of new housing on greenfield sites ... will ... damage biodiversity' (DETR, 1999: 7). The assumption is that biodiversity occurs on rural greenfield sites, but not on urban brownfield sites. This assumption is not supported by ecological science, which indicates that vast tracts of arable monoculture in the countryside support very little biodiversity, whereas urban areas are often highly biodiverse due to the variety of un-managed habitats provided by brownfield sites (Shirley and Box, 1998) (Figure 5.4). Within the context of urban liveability, the amenity value of urban open space within easy reach of a massive population is vastly higher than that of a field of wheat in East Anglia. Brownfield development represents a key arena in which these tensions within the notion of sustainability are worked out on the ground, as the next case study shows.

Case study: Selly Oak

The residential area of Selly Oak lies 4 km to the south of Birmingham city centre, 6 km from the city periphery. Birmingham City Council adopted a Local Action Plan for the area in 2001 to provide a framework for its sustainable development. The specific goals of the plan were to re-vitalise the shopping centre, attract quality investment to underused sites, increase public transport and cycling provision, enhance the environment for residents, and improve conditions for major land users. The goals reflect national priorities for sustainable development set out at the time, seeking to attract investment in order to improve the quality of an area.

The development plans for the area hinged on the redevelopment of a brownfield site known as Vincent Drive. Vincent Drive occupied approximately 30 hectares of floodplain

Figure 5.4   A river corridor on the Vincent Drive brownfield site in Birmingham. The area adjacent to the river is made up of important wetland habitat, which supports a range of rare species, such as the water vole. It is recognised as being ecologically valuable at both the regional and national levels, and indicates the sometimes surprising levels of biodiversity found on brownfield sites.

to the north and south of the Bourn Brook, and was the largest remaining semi-natural green space in south Birmingham (Figure 5.5). The northern half of the site was acquired by the Cadbury Trust (the landholding/charity arm of local philanthropist George Cadbury's chocolate empire) in the nineteenth century and sold to Birmingham City Council for £100,000 in 1926, for the development of the Queen Elizabeth Hospital. The southern half of the site, Battery Park, was home to a number of small industrial ventures that left behind substantial waste tips. By 1950 the site was distinguishable as a discrete green space, enclosed by Victorian terraced housing to the south and east, post-war housing to the west, and the Queen Elizabeth Hospital to the north.

The legacy of the old gardens, lack of management and varying drainage created an array of habitats on Vincent Drive. The remnant deer park land and pasture was over 1,000 years old, with no records existing of any ploughing having occurred. The Bourn Brook was characterised by well-developed woodland, with wetland habitat along its eastern stretch, and the linear features of remnant hedges were still identifiable. The Vincent Drive site was designated a Grade C Site of Importance for Nature Conservation (SINC) with parts of the site considered to be at the level of a Site of Special Scientific Interest.

As national policy shifted towards urban regeneration and brownfield development in the 1990s, a key task of the first Unitary Development Plan drawn up for the city in 1990 was to identify large sites for potential development within the city. Vincent Drive was earmarked for two major developments. Proposals from the Birmingham NHS Trust involved building a £300m hospital on Vincent Drive, which will cover approximately two-fifths of the site and provide 1,009 beds. Although the planning proposal was only officially submitted in 1999, the Regional Health Authority adopted a major hospital building programme

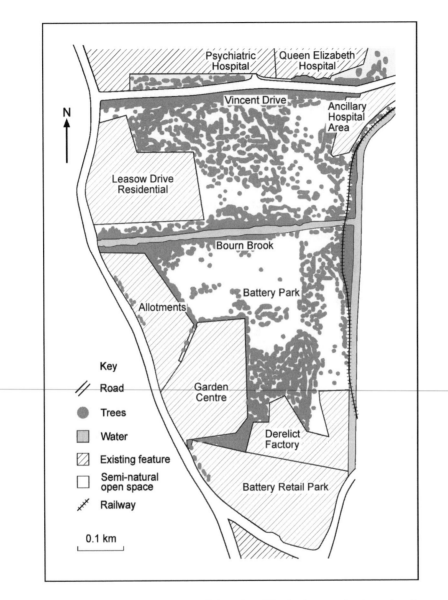

Figure 5.5 Land uses at the Vincent Drive site, Birmingham, prior to redevelopment. Note the very large area of semi-natural open space in the centre of the site, and the variety of land uses around the edges that have accumulated over time.

for the city ten years earlier, outlining the closure of some 20 hospitals and the centralisation of hospital services on four mega-hospital sites, as part of a regional health strategy. The developer's environmental consultant stated that 'the proposals are a positive move towards sustainability in that the new hospital would be provided on a brownfield site' (Babtie, 2000: i). Sainsbury's supermarket chain agreed to finance a new link road traversing the floodplain, canal and railway, and alterations to major road junctions at either end in return for development rights over the southern half of the site. Their proposals for a mega-Homebase DIY store and supermarket involve the complete reclamation of Battery Park, including the decontamination of waste tips and soil patches, removal of the derelict buildings and levelling of the site.

A series of local groups contested the plans. The local Wildlife Trust highlighted the loss of valuable habitat, and questioned the decision to site a major development in a floodplain. Local resident groups protested that the extra traffic the scheme would generate would overload an already heavily congested local road network. Politicians questioned how a regional hospital and retail development intended to serve the entire of the southern half of the city were contributing to local sustainability as part of the local action plan. Local business groups protested that far from revitalising the local shopping centre, the new development would shift the retail centre of gravity away from the high street.

The final development proposals clearly attempt to juggle economic development and open space provision in line with government policy. It was suggested that the loss of habitat could be offset by creating new habitats elsewhere, but the destruction of the area's largest natural resource is not offset by plans for similarly sized or equivalent replacement space in the locality. A key environmental element involved the enhancement of the Bourn Brook as a wildlife corridor, which would function as a recreational feature, with walkways, benches and landscaping (Birmingham City Council, 2002).

Vincent Drive highlights key tensions within the brownfield agenda (Evans, 2007). It is a highly diverse site, with areas of high ecological worth and areas of serious contamination. This makes it difficult to apply an overall judgement as to the development's environmental sustainability. In de-contaminating the waste tips, the development is improving the environment, but in building on a flood plain on a large semi-natural habitat it is undoubtedly destroying an environmental asset. As some studies have suggested, this problem may be partly related to the lack of urban ecological know-how in the planning and development sector, which tends to have difficulty in representing these habitats in the decision-making system (Harrison and Davies, 2002). Comparing Figures 5.5 and 5.6 gives a clear indication of the trade-off between development and enhancement of open space. The key environmental trade-off was achieved by the planning department requesting that the developers retain and enhance the habitat alongside the Bourn Brook, and improve public access to this area, in return for permission to develop the bulk of the site.

The outcome highlights the shift in policy away from speaking purely about environmental aspects, such as biodiversity conservation and air quality, to focusing on the provision of quality open space that is publicly accessible. In some ways there is a shift in practice from an environmental agenda towards a liveability agenda. The case study also highlights the multiple understandings of what sustainability actually means. Planners, developers and residents all have different interpretations based upon different priorities. A regeneration scheme that is sustainable at the city scale may not be sustainable at the local scale, while what may enhance the economic vitality of an area may damage it environmentally. Quite often it is argued that economic interests override other factors, as powerful interests exploit the ambiguity of what sustainability actually means to prioritise their own interests. While a number of studies have argued that this is indeed the case, one of the strengths of sustainability as a rallying point for partnerships is precisely that it is flexible and open to interpretation (for example, Evans and Jones (2007) have explored how common meanings of sustainability are established in regeneration schemes).

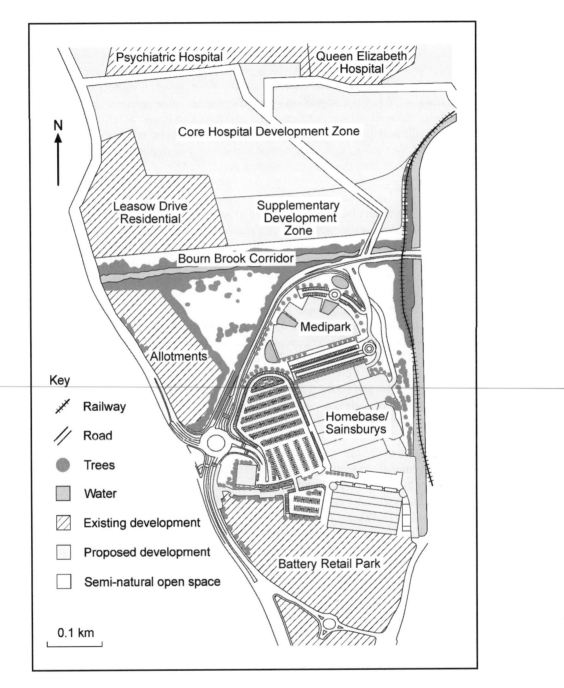

Figure 5.6   Land uses at the Vincent Drive site following development. The majority of the greenspace has been lost in the new development, with the retention of the river corridor the only major exception. Note the link road that opens up the middle of the site, paid for by Sainsbury's in return for development rights over the southern half.

Construction and the environment

Once the planners have reached a decision and the overall design for a site has been agreed, the actual job of building a development falls to construction companies. The construction industry has a massive impact upon the environmental aspects of regeneration projects. As argued in the first section of this chapter, the state cannot deliver the goals of sustainability alone, and the construction industry plays an important role in how a development is built and the specification to which it is built. In order to understand the environmental dimension of regeneration it is necessary to understand the policy drivers and practical workings of the construction industry.

Sir John Egan's task force report on the industry, *Rethinking Construction*, set out the key challenges and opportunities for the construction sector in the UK (Construction Task Force, 1998). The report focused on the need to improve the skill-base of the construction sector, and the need for the sector to work more closely in partnership with clients. The focus on the skill-base was not surprising given the challenges presented by the sustainability agenda, such as new technologies and integrated development, in the face of the progressive move away from a dependence on specialist labour on building sites. It was also reminiscent of similar issues concerning the lack of skilled workers that became apparent in the 1960s building boom.

The call for transparent partnership between the construction industry and other organisations involved in the regeneration process was a more innovative emphasis. Regeneration studies tend to focus on planners, architects, land agents and developers, forgetting that the construction industry is at the 'sharp-end', actually laying the bricks, pouring the concrete, and so on. By getting the industry involved, Egan hoped to get away from the traditional 'tendering' system, whereby a building or development would be designed, put out to tender and contracted to the construction firm who promised to build it in the specified time for the least money. This form of competitive tendering tended to lead constructors to assume a very short-term mentality, cutting corners in order to meet costs. Egan's emphasis on longer-term partnerships, engaging constructors at a far earlier stage in order to find the best method to deliver the product, would not only improve quality through the constructor having more of a stake in the development, but also efficiency through joined-up thinking.

Since the Egan report was published, an organisation called Constructing Excellence has been established to draw together government, the construction industry and major clients in order to identify and disseminate best practice. Constructing Excellence identified six areas in which the construction industry is central to the delivery of many of the government's policies for sustainable development:

- the regeneration of housing, particularly to revitalise town centres;
- planning communities to reduce car use;
- the protection of the countryside;
- minimising mineral extraction;
- using energy and water more efficiently; and
- the provision of training through schemes such as Welfare to Work and the New Deal.

The first four of these areas revolve around the development of brownfield land in urban areas.

Progress in the sector was reviewed in *The UK Construction Industry: Towards More Sustainable Construction*, prepared by the Sustainable Construction Task Group in 2003. The report concluded that while some companies have made significant progress, most have made no or limited numbers of steps towards more sustainable solutions. A key issue concerned information overload and the confusing array of initiatives. This is partly due to lack of coordination at government level, where construction is spread across many departments, and the failure of information to be focused on the business benefits of actions towards sustainability. The biggest barrier to establishing closer links is probably the institutional separation between CLG, which deals with planning, and the DBERR, which deals with the construction industry. Institutional separation is compounded by the divergence between the goals and language of integration associated with the public realm and the technical language of the private sector (Moore and Rydin, 2008).

In terms of building practices, the industry tends to react to environmental legislation. Building regulations are an important driver of construction practices. The Building Regulations set standards for the design and construction of buildings primarily to ensure the safety and health of people in or around those buildings, but also for energy conservation and access to buildings. The government also sets energy efficiency standards for new-build homes, measured using Standard Assessment Procedure (SAP) ratings. Interestingly, the government has tended to encourage sustainable construction more than legislate for it, preferring demonstration projects like BedZED in south London to legal regulations. The Beddington Zero (Fossil) Energy Development (BedZED) is a mixed-use scheme of 82 homes and 3000m^2 of commercial space in south London that has been developed by the Peabody Trust housing association (Figure 5.7). The scheme is famous for making use of environmental building technologies to reduce energy and water uses, and by providing accommodation that doubles up as work and living space. The BedZED model has become quite influential and has been applied to designs for a large urban settlement in China.

The reliance on exemplars rather than legislation may in part be a response to the construction lobby, which is generally averse to any tightening of building regulations on cost grounds. While it can be criticised for failing to force builders to become more sustainable, it has led to an interesting situation whereby sustainability in the construction sector is often equated with innovation (Moore and Rydin, 2008), and more responsibility is placed upon local authorities to negotiate the kinds of development that they want.

The Building Research Establishment (BRE) has developed an EcoHomes standard that assesses the environmental friendliness of a development. It considers energy, transport, pollution, materials, water, land use and ecology, and health and well-being. The scores can be aggregated to give a scheme an overall rating, from 1 (pass) to 4 (excellent), that can be used as a marketing tool to attract buyers. These standards are critical in making developments more environmentally sustainable, by reducing resource use and waste, but also play an important part in promoting social sustainability. For example, more efficient heating helps combat fuel poverty and improves human health. Housing associations are now required to achieve at least a pass at the EcoHomes standards in order to qualify for a subsidy from the Housing Corporation.

How have these policies impacted upon actual construction practice? The construction industry in the UK produced 91m tonnes of waste in 2003, of which 29m tonnes went to landfill. This accounted for 32% of all waste sent to landfill in the UK. This is not just an environmental issue, but has serious financial implications. It costs approximately

Figure 5.7   BedZED, the Beddington Zero Energy Development, is an architecturally striking scheme with echoes of the Victorian terraced street. It sets the benchmark for environmentally friendly urban development.

£24 per tonne to send material to landfill, in addition to the associated haulage costs, and this figure is set to double by 2010. In the case of material from contaminated land, there are only a few waste sites in the country that can accept it, further increasing costs of disposal. It has thus become a priority within the industry to find ways to reduce the amount of waste that is removed from sites. This can involve mixing contaminated soil with uncontaminated soil on larger sites in order to bring the average levels of contamination below the legal thresholds. Rubble from demolished buildings is sorted in order to recover steel, and concrete is crushed and graded for aggregate.

Similarly, green building technologies are being promoted not only because they are environmentally friendly, but because they are cost-effective in the longer term. The Royal Institute of Chartered Surveyors' collaborative report, *Green Value: Green Buildings, Growing Assets* (2005), draws upon Canadian expertise to suggest that eco-friendly buildings have economic advantages which developers are not taking advantage of. So, for example, green buildings have marketing potential to domestic and commercial customers, are associated with better health and have lower running costs. The Egan report (Construction Task Force, 1998) promoted the life-cycle cost analysis approach, which judges a technology on its construction *and* running costs. So, for example, installing better insulation will cost more during construction, but may prove cheaper if lower heating costs are taken into account over the life of the building. This approach complements Egan's emphasis upon partnership and longer-term engagement between construction companies and their products. Developers who wish to sell a development immediately will not benefit from this approach, whereas developers who have a long-term stake in the buildings' profitability will benefit from maximising rental value through attracting happy tenants.

Traditional accounting practices have undermined this approach by separating capital costs (initial costs of construction) from operating costs (heating, water, lighting, etc.). However, new modes of owner-occupation, such as Private Finance Initiatives (PFIs) (see Chapter 4), have encouraged the use of environmentally friendly technologies, as the PFI company not only meets the construction costs, but also absorbs the running costs for at least 25 years. The most recent government report on sustainable construction (DTI, 2006a) emphasises the need to standardise techniques such as life-cycle analysis within the industry, a conclusion which echoes Guy and Shove's (2000) analysis of knowledge in the realm of energy-efficient buildings.

Other studies have indicated that the industry tends to overestimate the costs of environmentally friendly technologies. Bartlett and Howard (2000) have argued that quantity surveyors, who examine how much building plans will cost given certain materials and technologies, tend to overestimate the costs of installing energy-efficient technologies by between 5% and 15%. They argue that if technologies are integrated into the design and construction process at every stage, then they should add no more than 1% of the overall cost of the building. Often these savings are invisible, because costs incurred in one area will produce savings in another. For example, while energy-efficient windows will cost more, they will save money in other areas, for instance through less powerful heating equipment being required. Again, Bartlett and Howard's study resonates with the Egan recommendations, highlighting the lack of expertise within the industry in this area and the failure of designers to work collaboratively with different teams of construction engineers. The case study in the next section explores how sustainable technologies can be incorporated into regeneration developments.

Key points

- Regeneration in the UK seeks to preserve the countryside by re-using urban brownfield land for development and ensuring that new developments are environmentally friendly.
- Tensions between the logic of high-density development and the provision of a quality environment are often worked out at the development level.
- Construction has a massive impact on the environmental sustainability of regeneration projects, both in terms of the construction process and the technologies used, but rarely goes beyond legal and financial requirements.
- The Egan agenda pushed for a more highly skilled construction industry that works with developers and designers rather than simply providing the cheapest tender.

Masterplanning and sustainability

As the name suggests, masterplanning involves taking a strategic approach to the regeneration of an area. As such it tends to be seen as an ideal vehicle to achieve the holistic goals of sustainable development. Although increasingly common, there is no single accepted definition of what masterplanning is. CLG suggests that a masterplan should provide a vision, a strategic decision-making tool, and marketing elements. Building on these essential characteristics and those identified by the Urban Task Force, CABE (2004) summarise a masterplan as:

- visionary – raising aspirations and providing a vehicle for consensus-building and implementation;
- deliverable – taking into account likely implementation and delivery routes;
- fully integrated into the land-use planning system while developing a site to its full potential;
- flexible – a base for negotiation;
- the outcome of participation; and
- rethinking existing neighbourhoods and creating new ones.

CABE identifies three elements to a masterplan: the strategic framework, the spatial plan and the implementation plan. The strategic framework contains a statement of aims and objectives for physical regeneration over a large area of land and may consider a wider area than the development itself. The strategic framework establishes baseline information, relating to the physical, social, economic and political context, in order to match them to the aims and objectives of the client. This framework is then turned into a three-dimensional spatial plan, which presents proposals or aspirations (but not actual designs) for the development of buildings, street blocks, public spaces, streets and landscape. This element of the masterplan draws upon a range of representational strategies, using diagrams, land-use plans, models and images as well as words. The implementation plan addresses all aspects of delivery, such as the programme, risk, funding and **procurement**. The three stages are obviously not entirely separate. For example, factors affecting delivery, such as social, commercial, political and economic realities that will drive change and development, should be assessed during the preparation of the strategic framework. Masterplans are not intended to provide a rigid blueprint for development, but a fluid and evolving strategic plan that can be refined as more information becomes available.

Masterplans are often supported by new organisations as existing structures are not adequate to deliver long-term strategic regeneration. Historically, the UK government has set up a variety of limited-life organisations to deliver a masterplan, such as New Town corporations, urban development corporations, Housing Action Trusts and, most recently, urban regeneration companies (URCs). The masterplan forms a focal point around which to ensure buy-in from stakeholders and the community. Due to its strategic approach it can also be used as a means of improving local services, by generating dialogue between local service providers and users. It also acts as a focal point to attract public and private investment into areas. Initial public investment can be used to encourage private finance, as was the case in the Salford Quays development, where the local authority lent considerable financial support to the initial private developments. Similarly, the masterplan provides a coherent selling tool for local and regional bodies to attract national and EU funding. By planning how land packages are released over time, it is possible to maximise the economic benefit of regeneration in an area, as the price of later packages rises in line with the improvement of an area. Carefully planning the schedule of land release also enhances the sustainability of a regeneration project. Developers are more likely to incorporate infrastructure such as roads, water and environmentally friendly technologies if the land package on offer to them is economically attractive.

The masterplanning process can be seen as a product of current trends in British planning. It combines long-term partnership between all those involved in the regeneration of an area with an emphasis on the spatial plan. However, it has clear antecedents in British planning. For example, a comprehensive approach was adopted by government in their plans for New Towns such as Stevenage and Milton Keynes to address the post-war legacy of damaged homes and disrupted industrial infrastructure. The Neighbourhood Renewal Unit (NRU) contrasts the success of the New Town programme with the local authority council housing programmes of the 1960s and 1970s, when homes were built rapidly without community facilities in an attempt to meet housing needs. While it is unfair in some ways to contrast greenfield developments in relatively affluent areas with inner-city slum clearance programmes, it is a comparison that is made nonetheless. Masterplans are thus seen as providing a way to meet current housing needs while avoiding the errors and pitfalls of the large-scale building programmes of the 1960s and 1970s.

This chapter has highlighted some key challenges facing sustainable regeneration. It is apparent that social and environmental priorities often take a backseat to economic considerations in regeneration developments. There are also difficulties in achieving positive changes beyond the spatial boundaries of specific developments. The Egan Report (Construction Task Force, 1998) highlighted the need for longer-term partnerships between constructors, developers and planners in order to deliver sustainable developments. It is within this context that masterplanning is seen as a particularly useful tool to achieve the goals of sustainability. They are based upon stakeholder engagement and the empowering of local communities through the strategic and partnering approaches to housing planning and development frameworks. This strategic approach helps planners to ensure that a set of developments creates mixed communities rather than islands of gentrification. Similarly, developments can be planned to respond to needs at the sub-regional level, rather than only considering the requirements of a particular scheme. More specifically, the plans can help local authorities meet floor targets for health, education, crime, unemployment, housing and liveability for deprived areas outlined in the government's Public Service Agreements (PSAs). These include bringing all social housing into a decent condition (PSA 7) and delivering the liveability agenda to improve public spaces and the quality of the built environment (PSA 8).

Case study: Rotherham

Rotherham is currently undergoing a major period of regeneration, covering both the city centre and areas of run-down housing that are subject to Pathfinder initiatives surrounding the city centre. Both the town centre and housing projects are being masterplanned in order to ensure that the strategic goals of sustainability are met.

Rotherham Renaissance is a 25-year masterplan to completely revitalise the town centre along the corridor of the River Don, which runs through the town. Investment of approximately £2bn is forecast from the public and private sector, and the masterplan aims to address a range of goals:

- Make the river and the canal a key part of the town's future
- Populate the town's centre by creating *good quality living*
- Place Rotherham within a *sustainable landscape* setting of the highest quality
- Put Rotherham at the centre of a *public transport* network
- Improve parts of major road infrastructure
- Make Forge Island a major new piece of the town centre
- Establish a new *civic focus* that not only promotes a more open and accessible type of governance but also embraces culture and the arts
- Demand the best in architecture, urban design and *public spaces* for Rotherham
- Improve *community access to health, education* and promote *social well-being*
- Create a broadly based, dynamic local economy with a vibrant town centre as its focus. (Rotherham Metropolitan Borough Council, 2007, emphasis added)

The rhetoric of sustainability clearly motivates this list, as the italicised items highlight, and the Rotherham Rennaissance has undoubtedly achieved a number of early wins. For example, the Moorgate Crofts Business Centre, opened in 2005, is an award-winning business incubation centre located along the regeneration corridor. The building utilises state-of-the-art environmental technologies, built from a mixture of traditional, modern and recycled materials, using geothermal heating and cooling, and having a 'green' roof covered with more 8,000 plants. The building provides 60 highly flexible small business units, which can be leased for short-term periods, and has dedicated space for start-ups, in order to encourage young people into business. Moorgate Crofts is owned and operated by Rotherham Investment and Development Office, part of Rotherham Metropolitan Borough Council, and shows how the economic priorities outlined in Chapter 4 can be united with environmental technologies.

While the Rotherham Renaissance masterplan adheres to a familiar blueprint, focusing on the opportunities of the River Don for riverside developments, and the creation of a new civic, cultural and arts quarter, featuring green public spaces and boulevards, it is interesting to note certain ways in which the ideas of sustainability are negotiated. For example, because the plan has been driven by the views of the town's residents, there is a major emphasis on the provision of more cheap parking in the centre through the building of multi-story car parks. Road improvements also seem to take precedence over public transport. The masterplan is also highly aspirational in nature, involving plenty of mission statements and 'visions' for the town, but not so many concrete planning guidelines or requirements. To some extent this is a function of the fact that the programme is in its early days, and many of the detailed elements of the masterplanning process are being contracted out to private consultants. For example, Roger Evans Associates were commissioned by Rotherham Metropolitan Borough Council to prepare 'design codes' for Rotherham Town Centre and River Corridor as part of CABE's design code pilot (design coding will be discussed in Chapter 6).

Turning to the regeneration of residential areas, Rotherham East is a housing area close to the town centre, characterised by a high number of houses below the government's benchmark for housing standards and low levels of private home ownership. The area also has high rates of unemployment and low levels of educational achievements. The Housing Market Renewal schemes aim to replace obsolete housing with modern sustainable accommodation, through demolition, refurbishment and new building over a

10–15 year timeframe. Shillam and Smith have been commissioned as independent consultants to develop a strategic masterplan for how the area may look in the next 15 years. The masterplan also fulfils bureaucratic functions, forming the basis for the local planning authority's development framework, and being used to provide an interim report to the government who then allocate funding for housing renewal. The masterplan focuses on strategies for improved housing quality, but also ensures that the other essential requirements of sustainable communities are addressed, such as quality services, good design and clean, safe, healthy and attractive environments. Major schemes are proposed to develop the landscape as a green lung and improve the neighbourhood centres within the area.

There is no doubt that masterplanning is well suited to delivering more sustainable developments. Its long-term, strategic partnership approach resonates with the fundamental tenets of sustainable development. It is not, however, a panacea for the challenges of creating more sustainable cities. Trade-offs still have to be made, and research has indicated that economic factors still assume overriding importance within regeneration partnerships, with a tendency for actors to revert to tried-and-tested strategies in the face of uncertainties over how to implement sustainability (Donovan et al., 2005). This can result in a lack of imagination in terms of proposed developments, with buildings and layouts almost indistinguishable from generic regeneration schemes and a reticence on the part of developers to innovate with sustainable technologies unless forced to. The process of negotiation between different stakeholders in the masterplanning process is highly politicised and, as Rydin et al. (2003) have argued, sustainability can get lost amid pre-existing conflicts between antagonistic groups.

Finally, it must be remembered that masterplanning constitutes a form of area-based regeneration. However inclusive a process it may be, it still involves the comprehensive redevelopment of a bounded area. Large land packages are often needed to allow strategic planning and are seen as more appealing to developers; this often involves compulsorily purchasing land from people who are unwilling to sell. This can lead to tensions with the pre-existing populations, however small, as the area is represented as an 'empty space' awaiting development. Effacing local populations from an area's future in this way can throw doubt over the inclusivity of the masterplanning process, and raises similar questions concerning social sustainability that were explored in relation to Salford Quays.

Key points

- Masterplanning allows the strategic goals of sustainability to be addressed, helping planners meet government floor targets, and can help attract funding to an area.
- There are three elements to a masterplan: the strategic framework, the spatial plan and the implementation plan.
- They are not panaceas for sustainable development; trade-offs must still be made between different land uses and stakeholders.

Conclusions

Sustainability informs a number of major policies that frame urban regeneration, including Planning Policy Statement 1 and the Sustainable Communities Plan. There is no doubt that the integrated nature of urban regeneration makes it the ideal vehicle to deliver the holistic goals of sustainability. The long-term, large-scale approach of regeneration facilitates a strategic dimension of planning to incorporate social and environmental issues into schemes that may not exist in piecemeal developments. This is a vital part of the challenge to make cities attractive places in which to live, and can be seen as a key element of the wider regeneration agenda. In many ways, the idea of sustainable regeneration can be seen as the means by which the government is attempting to ensure that a balance is struck between the purely financial benefits of new developments for the private sector and longer-lasting benefits to local communities.

This chapter has explored the challenges involved with balancing economic, environmental and social factors, and has identified key tensions within the process. One of the most important arenas in which these tensions are expressed is through debates surrounding housebuilding. Housing provision lies at the core of the SCP, and Gordon Brown made housing provision the key policy at the start of his premiership, emphasising the need to increase the access of all sectors of society to affordable housing. This may mitigate the problems of gentrification that were highlighted by the Salford Quays example, as it represents a shift towards ensuring social equity for all social classes. But within this current debate, the housing minister has hinted that this may have to be at the expense of environmental considerations, refusing to deny claims that green belt regulations may be relaxed in order to achieve the new higher national targets for housebuilding. The notion of 'sustainability' does not offer a perfect predetermined solution, but is still in the process of being worked through in policy. Despite these difficulties it seems set to form the major approach to urban regeneration for the foreseeable future.

Further Reading

It is hard to place boundaries around the topic of sustainable regeneration, as it is an area that evolves rapidly and cuts across urban, environmental and social fields of inquiry. Girardet (2006) provides a thought-provoking consideration of the broader context of sustainability and urban regeneration. Owens and Cowell (2002) provide a thorough overview of the general challenges sustainability poses to the UK planning system. Barton (2000) and Gibbs (2002) provide interesting discussions of the role communities can play in achieving more sustainable urban development. Jenks and Dempsey (2005) explore how urban design and planning contributes to sustainability at a range of scales, while Edwards and Turrent (2000) consider the role of sustainable construction in housing. Raco and Henderson (2006) give an up-to-date account of brownfield planning and sustainable regeneration, and Cowan (2002) presents one of the few studies of urban masterplanning.

Barton, H. (ed.) (2000) *Sustainable Communities: The Potential for Eco-neighbourhoods* (Earthscan, London).

Cowan, R. (2002) *Urban Design Guidance: Urban Design Frameworks, Development Briefs and Master Plans* (Thomas Telford, London).

Edwards, B. and Turrent, D. (eds) (2000) *Sustainable Housing: Principles and Practice* (Spon Press, London).

Gibbs, D. (2002) 'Urban development and civil society: the role of communities in sustainable cities', *European Urban and Regional Studies*, 9: 350–351.

Girardet, H. (2006) *Creating Sustainable Cities* (Green Books, Totnes).

Jenks, M. and Dempsey, N. (eds) (2005) *Future Forms and Design for Sustainable Cities* (Architectural Press, Oxford).

Owens, S. and Cowell, R. (2002) *Land and Limits: Interpreting Sustainable Development within the Planning System* (Routledge, London).

Raco, M. and Henderson, S. (2006) 'Sustainable planning and the brownfield development process in the United Kingdom', *Local Environment*, 11: 499–513.

6 Design and Cultural Regeneration

Overview

This chapter examines the increasing emphasis on producing high-quality urban design and the value of promoting culture to foster regeneration.

- *The urban design agenda*: explores how notions of 'good' urban design have infiltrated urban regeneration policy, including the role of the Urban Task Force report and the Commission for Architecture and the Built Environment. The influence of new urbanism on UK urban design is also examined.
- *Cultural regeneration*: explores the broader notion of culture-led development through case studies of waterfront redevelopment, the re-use of historic buildings, cultural anchors and signature buildings, cultural quarters, sub-cultural quarters, Liverpool as the European Capital of Culture 2008 and sports-led regeneration.

Introduction

The meaning of the word 'culture' is somewhat ambiguous. Individuals and communities can be labelled as belonging to a particular cultural category and the term can also be used to signify the products of those individuals and communities, whether these be material objects or performances. Hence culture allows us to think about symbolism and the image of cities as well as a mindset that drives decision-making in particular directions. In urban regeneration the term 'culture' tends to be used rather uncritically and is applied to a whole range of issues, from design and architecture through artistic works and sporting events to a more general sense of creativity and the knowledge economy.

Contemporary regeneration places a great emphasis on 'high-quality' design. This too has a somewhat ambiguous meaning, combining notions of both aesthetic and functional quality – the way things look and the way that they work. Assessments of aesthetics and even function are somewhat subjective, making 'design' difficult to measure and audit. Design quality is frequently linked to the broader idea of cultural regeneration, with aesthetic and functional design at the heart of 'good' urban regeneration. An obvious manifestation of this is where new flagship cultural resources – whether this be an art gallery or a football stadium – are seen as having anchoring qualities for regeneration programmes, with an architecturally impressive new building attracting private investment to a previously unfashionable area. It must be noted, however, that securing an iconic building for a site does not guarantee that the cultural resource within it will be a success, nor that other regeneration activity will follow in its wake.

Notions of cultural regeneration are often seen as a key strategy in making cities economically competitive. This brings us back to Richard Florida's (2002) argument, discussed in Chapter 4, about the need to attract members of the 'creative class' working in the knowledge economy. Flagship buildings and cultural resources are a part of this, but also the idea of fostering cultural clusters or cultural quarters, where creative people can work in close proximity and develop the kinds of 'soft' personal networks that are seen as crucial for innovation in the knowledge economy. This raises questions about *which* kinds of culture are incubated within these cultural quarters and whether policies to encourage economically 'valuable' cultural clustering risk shutting out more marginal or experimental cultures.

The urban design agenda

Urban design has been around as long as settlements themselves, from the simple defensive ditch to the Greek *agora*, to the medieval burgage system and onwards through history. It was not until the nineteenth century that architecture emerged as a specific discipline, with the idea of planning as a regulatory mechanism subsequently emerging by the early twentieth century. Urban design as a distinct concept is a more recent arrival and seeks a synthesis between the architecture and planning disciplines, looking at the aesthetics and functioning of towns and cities at a variety of spatial scales. In part, the notion of urban design can be seen as having come out of a frustration about a lack of integration between individual buildings and the townscape as a whole. It tends to be taken as axiomatic in policy discourse that good urban design is a prerequisite for good urban regeneration.

David Bell and Mark Jayne (2003) have taken issue with the notion that high-quality design really makes a significant difference to urban regeneration strategies. Indeed, they argue that the meaning of 'design-led' regeneration is fundamentally ambiguous, in spite of attempts being made to assess the impact of design activity on redevelopment projects. The phrase is most clearly associated with things such as flagship buildings and improvements to the **public realm** (squares, fountains, landscaping, etc.) which are then tied into more traditional place-marketing strategies in order to attract economic activity to the area (see Chapter 4). The new cafés, restaurants and other service facilities that are attracted have become the signifiers of a post-industrial urban economy. In a sense, this is an attempt to build the kind of urban environment that will attract the 'creative class' of designers, media people, policy-makers, ICT workers and so forth, who are believed to drive the knowledge economy.

Urban design policy

The report of the Urban Task Force (1999) was a key document in cementing the idea of design-led regeneration in policy discourse. The report itself is a vision statement rather than a policy document, and it should come as no surprise that, with internationally renowned architect Richard Rogers leading the Task Force, high-quality urban design should be given a great deal of emphasis. Nonetheless, some of the ideas about design found their way into the subsequent Urban White Paper (DETR, 2000). In part, the emphasis on good urban design can be seen as a reaction against the perceived failings of post-war urban design, with its concrete canyons and crime-ridden estates. This kind of discourse makes an implicit link between environment and human behaviour in a way that many social scientists would be rather uncomfortable with. As an aside, it should also be noted that some of the design features of the post-war period, for example the notorious 'rabbit warren' Radburn layout, were seen as best practice at the time.

Six years on from the original Urban Task Force report, a follow-up document was produced, this time without being commissioned or supported by the government. In the introduction, Rogers explicitly criticised the continuing poor quality in urban design: 'Few well-designed integrated urban projects stand out as international exemplars of sustainable communities, despite public investment in new housing' (Urban Task Force, 2005: 3). Here an explicit link is made between high-quality design and the creation of a sustainable community. Commentators generally have lamented the quality of much urban design since the original Urban Task Force report, although whereas post-war design is often seen as monstrous, post-1999 design is merely bland and characterless. Whether the mere fact of living on a generic-looking housing estate inhibits people's ability to form a sustainable community (however defined) is moot, but there can be no doubt that few standard housing schemes built over the last decade have excited those with an interest in urban design.

The concept of **mixed development**, heavily promoted by the Urban Task Force report, has become somewhat of a shibboleth in contemporary urban design. The vision seems to be one where central cities are filled with three- to five-storey apartment blocks – preferably built on brownfield land – containing live/work spaces, where people walk to access their local shops and services. By cutting down on travel time and carbon dioxide emissions, mixing places of business and residential accommodation is seen as uncritically a good thing – much as it was taken as read during the post-war reconstruction that it was a good thing to rigidly separate residential accommodation from places of work. It would be unfair to characterise the Urban Task Force report as completely ignoring suburban, family housing, but there is a palpable enthusiasm for a somewhat **metrocentric** model of living, perhaps overly biased towards the needs of young, childless professional couples. This tension between city-centre modes of regeneration and suburbia is discussed further in Chapter 7.

The Commission for Architecture and the Built Environment (CABE), established in 1999, expanded the urban design remit of its predecessor, the Royal Fine Art Commission. CABE has defined the objectives of urban design as being sevenfold:

- Character: a place with its own identity.
- Continuity and enclosure: a place where public and private spaces are clearly distinguished.
- Quality of the public realm: a place with attractive and successful outdoor areas.

- Ease of movement: a place that is easy to get to and move through.
- Legibility: a place that has a clear image and is easy to understand.
- Adaptability: a place that can change easily.
- Diversity: a place with variety and choice. (CABE, 2000: 15)

These are quite interesting in that they indicate the breadth of the concept, but also because they can be seen as reacting to some of the perceived failures of modern architecture and planning as it was played out in post-war Britain. Issues of delineating public and private space, for example, were a major criticism of modern planning in that it created spaces where there was no sense of 'ownership', which thus went unregulated by the community. The architectural critic Jane Jacobs (1961) referred to this regulatory effect as the 'eyes on the street', an idea which was later formalised into the concept of defensible space (Newman, 1973). Contemporary urban design frowns upon spaces where public and private are not clearly divided, something reinforced through the advice given to designers by Police Architectural Liaison Officers, who widely employ the concept of defensible space.

The anti-modernist stance is also clearly seen in the notion of legibility, which developed out of the work of Kevin Lynch (1960), who criticised urban spaces where movement and function were not clear. One only has to think of the muddle of underpasses, changes in level and blind corners that typified post-war shopping precincts in the UK to understand this emphasis on legibility in contemporary urban design. In the rhetoric of legibility, the subtleties of Lynch's argument are sometimes lost, but there has been a renewed emphasis on lines of sight in masterplanning, attempting to render spaces more readily understandable to people entering them for the first time. Similarly, Lynch's idea of 'imageability' explores the characteristics of areas that are particularly vivid and distinct – as well as pleasurable – in people's mental maps of the city. This notion of visual clues which give a city coherence to the person moving around it clearly reacts against the blandness of many modernist city developments. Again, the general idea of imageability has been embraced, in rhetoric at least, by those interested in contemporary urban design.

CABE (2005b) responded enthusiastically to the reforms in the Planning and Compulsory Purchase Act 2004, particularly replacing local and unitary development plans with smaller scale and more flexible local development frameworks (LDFs). The belief is that LDFs will deliver high-quality buildings and spaces by laying down spatial design policies at a variety of different scales. Again, however, the idea of high-quality 'design' is used uncritically and without quite stating what this means. At least, however, the move towards LDFs may help overcome a traditional UK problem of seeing 'design' as a site-specific concern, rather than looking at how good design fits into the wider urban environment. LDFs may, therefore, become something akin to a good redevelopment masterplan.

LDFs key into another planning innovation being championed by CABE. From 2004 to 2006, along with the then Office of the Deputy Prime Minister and English Partnerships, CABE piloted a series of 'design codes' in seven areas. The idea of design coding is fundamentally inspired by new urbanism (see below) and designates a series of design specifications for a redevelopment area, with varying degrees of detail at different spatial scales. These specifications might include, for example, height of building, storey height, set backs from the street, overall street/frontage patterns, guidance on materials, textures and colours, sometimes even specific guidance on detailing of buildings and public realm infrastructure. The overall intention of these codes is to make design quality auditable, such that where developers stick within the design code they should have an almost automatic right to develop on the site that has those guidelines attached to it. The presumption is that if developers stick to the code, then good designs for the area should automatically follow. Design codes are used widely and successfully on the continent, especially in Germany, although Richard Rogers has been somewhat scathing about the aesthetic results of design code-led urban planning. This said, there is general agreement that good design codes can at least stop the worst architectural abuses and can inspire genuinely diverse and interesting architecture. The team led by the Bartlett School of Planning, which evaluated design coding, gave it favourable reports and it was subsequently integrated into the new *Planning Policy Statement 3: Housing* (CLG, 2006c), making it now a fairly standard part of the English planning system.

New urbanism

Proponents of design coding in the UK have been inspired by the work of the new urbanist movement. Although the term 'new urbanism' has had less exposure in the UK than the United States, many of its key tenets have been absorbed into UK policy, in particular the rhetoric of mixed development, diversity and compact walkable cities (Talen, 1999). In the USA, new urbanism is frequently entangled with the debate over 'smart growth'. Where new urbanism as a movement was developed by architects and planners, smart growth was developed by environmentalists and policy planners. As a result, the terms are not exactly interchangeable, although both talk about mixed uses and walkability. New urbanism stresses design elements where smart growth is driven by considerations of economic development and environmental sustainability.

Some social scientists have been a little uncomfortable with the kinds environmental determinist claims made by new urbanists that 'good' design can promote 'good' communities. Equally, however, new urbanist practitioners complain that social scientists do not engage with the practicalities of delivering urban regeneration on the ground and the importance of good design for achieving this. New urbanists, architecturally, have leanings towards the neo-vernacular, which means highly qualified architects self-consciously producing updated versions of the historic building styles that were developed locally by non-architects. The adherence to the neo-vernacular is not universal, however. Indeed, the idea of design coding was developed in the new urbanist settlement of Seaside in Florida, famously the location for Peter Weir's film *The Truman Show* (1998). The film did, however, carefully pick locations in the town where neo-vernacular design predominated to generate a particular vision of an idealised Americana. Less traditional designs were employed elsewhere in Seaside, with design coding used to set general parameters to give an overall coherence to the town, but with variation permitted within those parameters.

The most prominent example overtly applying new urbanist concepts in the UK is Poundbury in Dorset, which was designed by Luxembourg architect Léon Krier to a commission by HRH the Prince of Wales. Work began on Poundbury in 1989 and helped to cement some of the principles which are now in the policy mainstream, particularly mixing

Figure 6.1   Poundbury divides opinion between those who like its village character and those who are uncomfortable with its fake historicity. Narrow winding streets lead to Brownsword Hall, which sits at the heart of the 1990s development and hosts local events.

residential accommodation and workplaces, high-density walkable urban areas and, in theory at least, mixed demographic and community groups. Stylistically, it's a 'myth-mash' of fake Georgian and fake medieval (Figure 6.1). The architectural critic Jonathan Glancey (2004) has been somewhat cutting about how the rhetoric has been played out, noting that the 'mixed uses' include things like equine vets, a chocolate shop and other rather bourgeois services.

Poundbury does make the car subservient to the pedestrian, tucked away in separate car parks with many roads entirely traffic free. Though low rise, it is a high-density development, with limited space around the houses. The socially rented and private housing is indistinguishable and inter-mixed, but as property prices have risen rapidly, it has become rather an expensive place to live. Poundbury's defenders would argue that the rise in property prices proves that the design works well and is successful because it delivers what people want from an urban settlement. While one may criticise the pastiche fake historicism on *aesthetic* grounds, in more general terms, there is no doubt that the design principles have been influential. The same principles are now being applied to a follow-up project by the Prince, with a planned extension to newly fashionable Newquay, the so-called 'Surfbury' development. Once again, design coding and the principles of walkable, mixed developments are at the fore; the difference is that where Poundbury was somewhat experimental, the ideas it pioneered are now at the heart of urban policy.

Key points

- Contemporary urban design is a synthesis of architecture and planning which examines the coherence of development at the area scale.
- The meaning of 'good' design is somewhat ambiguous, containing notions of both aesthetic and functional quality.
- There is a strong anti-modernist reaction in debates over contemporary design quality, with much of post-war urban design held up as worst practice.
- Key policy innovations in urban design are local development frameworks, which are akin to a development masterplan, and design coding, which gives developers a near-automatic right to build if they stay within specified design guidelines.
- New urbanist notions of walkability, mixed development and design coherence have become major tenets of urban design policy.

Cultural regeneration

Perhaps the most important factor driving the move towards regeneration through culture has been the development of the post-industrial economy described in Chapter 4. The key impacts of this have been twofold: first, very large, formerly industrial areas within cities, having fallen into disuse, have provided major redevelopment opportunities; secondly, the change to a more knowledge/services-based economy has created a broader cultural change within society, with different types of employment driving a demand for different types of work/leisure/residential space.

Waterfront redevelopment

The redevelopment of former industrial waterfronts is a classic example of how structural changes in the global economy have given rise to regeneration opportunities. Waterfront redevelopment has now become so common that it is almost the cliché of regeneration that where once men sweated in docks and shipyards, we now have middle-class professionals sipping cappuccinos in pavement cafés and living in converted loft apartments. The post-war move towards containerisation led to bigger ships needing deep water port facilities, which made many older docks in inner cities redundant. This created an opportunity to transform the image of cities, putting dockland areas to new uses and creating whole new quarters of the city. This has transformed the cultural identity of cities along with their physical appearance.

Waterfront redevelopment is not a particularly new phenomenon. In the late 1960s the city of Baltimore, with financial assistance from the United States federal government, began regenerating its Inner Harbor district. Pollution was cleaned up, historic buildings restored, parkland replaced dockland and through the 1970s the area became the location for various cultural events. This culture-based regeneration continued with the Inner

Harbor becoming home to the National Aquarium and Maryland Science Centre alongside Harborplace, a large retail and leisure complex. An abandoned, polluted space was transformed into a highly attractive quarter right at the heart of the city, helping to drive the broader cultural transformation and economic redevelopment of the city centre.

The model developed in Baltimore has been copied across the post-industrial world. Indeed, it is so successful that in Amsterdam entirely new islands have been built in the IJ river to accommodate the demand for this kind of development. In the UK, there has been two distinct phases of waterfront redevelopment, the first through the 1980s until the property crash of the early 1990s, the second picking up through the late 1990s and progressing very quickly from 2000. It is important to note that when Britain's industrial waterfronts were in use and when they were derelict, it was actually quite difficult to get access to the waterfront. People could not simply walk into a dockyard to have a stroll by the river. Indeed, no one would have wanted to promenade along the waterfront given the high levels of pollution suffered by many of Britain's working rivers. Through the 1980s, the decline of British industry and new regulations on discharges, combined with better sewage treatment, meant that water quality began to improve dramatically such that, for example, by the early 1990s salmon had returned to the once toxic River Mersey.

Waterfront regeneration has involved giving people access to the waterfronts where none existed before to compliment improvements in the marine environment. Improved water quality reinforces the innate attraction of waterfront sites, with the aesthetic appeal of reflections on the water, boats going by, leisure activities, and so on. This in turn can attract commercial attention. The pioneers of waterfront development in the UK were not private developers, however, but urban development corporations (UDCs). These Thatcherite instruments of central government intervention in acutely deprived areas had perhaps their most notable successes with Canary Wharf in London and the Albert Dock in Liverpool.

Canary Wharf is today a hugely wealthy office district built in what in the early 1980s was a largely abandoned area. Very little of the historic built fabric was retained in this project, with a focus very much on new high-rise offices and, subsequently, apartments. In terms of innovation, the work of the Merseyside Development Corporation in reviving the Albert Dock was perhaps more interesting. The architecturally valuable historic dock buildings could not be demolished and so were redeveloped. The ground storey was a mix of retail and cultural uses – including the Merseyside Maritime Museum and, later, the art gallery Tate Liverpool – with the upper storeys being a mix of offices and loft-style apartments (Figure 6.2). Granada Television also located offices and a studio in the redeveloped site (media companies are often perceived as the ultimate seal of approval for a cultural regeneration). The mix of uses at the Albert Dock was highly innovative for the UK at the time, with some elements of the scheme being quite high risk for a city with a depressed property market. Nonetheless, the mixing of uses insulated the Albert Dock from the slump in the office property market in the early 1990s, which saw the newly completed Canary Wharf partially empty and its owners filing for bankruptcy in 1992. After that early wobble, however, London Docklands has gone on to become a world hub for financial services.

On many levels one can depict these kinds of waterfront development as win-win. Polluted waterscapes have been cleaned up, historic buildings given a new lease of life,

Figure 6.2   Jesse Hartley's 1840s Albert Dock is now Grade 1 listed – the highest level of protection for historic buildings in England. In the 1980s a process of redevelopment produced a mix of offices, shops and apartments as well as cultural uses. The Albert Dock represented Liverpool's first major foray into urban regeneration and the city's latest scheme, a new arena and conference centre, has been built alongside it.

economic activity created from derelict areas and cultural uses given a new home. One criticism, however, is that many of these schemes do relatively little for the surrounding area. Taking the Docklands Light Railway across to Canary Wharf gave a view into a divided city, where acute poverty in crumbling social housing sat side by side with the extremes of wealth generated by the new financial district. Nonetheless, the dramatic visual impact of a whole new quarter of a city arising from derelict waterfront is exceedingly appealing and it is little wonder that this approach has proved so popular, especially with so many examples of similar successful projects elsewhere.

Today the Albert Dock is starting to look a little worn around the edges, but this has not discouraged a major new investment in the adjacent Kings Waterfront development. With no historic buildings to preserve here, an architecturally striking new complex has been created with an arena and major new exhibition space for the 2008 Capital of Culture (of which more below), along with the usual mix of residential and retail uses which are so familiar in any development. The model that Liverpool helped to pioneer is now so mainstream that this new project, though impressive in scale, does not seem particularly innovative, as a great many other interesting (and not so interesting) waterfront schemes have been carried out across the UK in the last 25 years. Gloucester, for example, which is not one of the 'usual suspects' in discussions of urban regeneration, has an ongoing programme redeveloping its inner dock areas. The completion of the Gloucester and Sharpness canal in 1827 allowed the massive expansion of the docks, becoming England's most significant inland port. As the docks fell out of use, they lay derelict until the mid-1980s when the City Council moved its main offices there as part of a major redevelopment scheme, although this only dealt with the interesting historic buildings closest to the city centre. More recently, a second wave of development saw a series of new schemes for the docks, including a mix of conversion and new build residential flats by developers Crest Nicholson.

The Gloucester Quays scheme received planning approval in 2004 as a partnership between British Waterways (who own the canals) and Peel Holdings, assisted by a newly established urban regeneration company (URC), Gloucester Heritage. The £200m scheme was granted final approval by the Government Office for the South-West (GO-SW) and the Department for Communities and Local Government (CLG) in June 2006. Covering 25 hectares of previously developed dockland, Gloucester Quays sits as part of a broader £1bn redevelopment of the city led by the URC. Interestingly, Gloucester Heritage is the only URC that places heritage as part of its mission, and the new development retained and refurbished some of the historic buildings while completely clearing the remainder of the site. Gloucester Quays mixes upscale residential, workshops, 'Designer Outlet' shopping, offices, a supermarket and Gloucester College of Further Education's new campus, all set around a cleaned-up canal and upgraded road infrastructure. The ten-year scheme will, therefore, produce some very significant changes to the urban landscape of inner Gloucester. There are some concerns about the impact this scheme might have on retail within the old shopping core of the city, meaning that it is not necessarily win-win, but it does represent an attempt to get maximum land-use value out of a large, semi-derelict area. It must be said, however, that the first of the newly built apartment developments are not particularly inspiring in terms of their architecture, although some attempts have been made to fit them into the profile of the historic dock buildings.

Figure 6.3   Postmodern architecture often borrows from past styles, but in this case, part of Birmingham's Orion Building, a section of terracotta Victoriana has been literally built into the new development. There has, however, been relatively little attempt made to integrate the old façade into the design of the new building.

Re-use of historic buildings

The Albert Dock is a good example of a particular kind of historic building, which has been culturally validated through being listed for preservation and given a new lease of life through redevelopment. Cleaned-up and recycled, such architecturally interesting buildings can lend character when sensitively integrated into a development programme. Indeed, English Heritage has promoted conservation-led regeneration as a model for protecting buildings and areas of specific historical interest. It is, however, considerably more expensive to re-use an old building than to create a new one from scratch, as the development process requires more ingenuity to fit new uses and new requirements for structural and environmental function into an existing superstructure. In addition, Value Added Tax is charged on building materials for refurbishment, but not for new build, giving an extra incentive not to retain older buildings.

One way around this problem is façadism. Essentially, the front of the building is kept, such that it looks the same to a passer by, but the remainder is demolished and a new building erected behind the façade. This can be particularly effective where properties appear in terraces, although it can look a little strange where an original façade is used to form the lower storeys of a much taller new building (Figure 6.3). There is, however, a broader philosophical point about façadism in that while it may preserve the aesthetics of the

Figure 6.4   Grainger Town in Newcastle retains the attractive streetscape of Richard Grainger's original development and combines it with a sensitive scheme of pedestrianisation. Some of the original urban fabric behind the façades has, however, been lost, which raises interesting questions about the trade-off between contemporary needs and historic value.

streetscape, it is fundamentally dishonest in that the functional logic of the historic building is destroyed. John Pendlebury (1999, 2002) has examined these issues at length through a study of Newcastle's Grainger Town district (Figure 6.4).

Speculative developer Richard Grainger was responsible for building a large portion of central Newcastle between the 1820s and 1840s, creating some rather wonderful streetscapes in the neo-classical style. While some of these buildings were destroyed as part of comprehensive redevelopments in the 1960s and 1970s, the inner core area around Grey Street and Grainger Street was made the subject of a conservation area. In the early 1990s the Grainger Town district was identified by the local council for a major redevelopment programme. Between 1997 and 2003 the Grainger Town Partnership, a coalition of the City Council, English Partnerships and English Heritage, led the regeneration of this area. The streetscape and façades were by and large retained in this process, but there was a tension between the demands of developers and their clients for large floorplate-type offices and the existing buildings which had smaller, less flexible spaces and in some cases were in a state of disrepair. A great many of these buildings were demolished, with the façades retained to front new, modern buildings with all the facilities demanded by contemporary businesses.

At the end of the project the Grainger Town streetscapes are by and large intact. Arguably, the cleaned and repaired façades, in combination with a scheme of pedestrianisation, have in fact significantly improved the aesthetics of the area. In many cases, however, the historic grain of the urban fabric has been lost, which raises a point about the kinds of culture which are valued in these schemes – in this case, aesthetics over substance perhaps. The counter-argument is that the Grainger Town district was declining in the early 1990s and the intervention has seen a significant improvement in its fortunes. Again, this reiterates a central point that regeneration schemes which overtly draw on notions of culture are not automatically win-win for the *existing* culture of an area; as with all regeneration, there are compromises to be made. Essentially, there is a tension here between culture as physical artefacts and culture as a way of life that is difficult to resolve.

The treatment of historic buildings in redevelopment is very dependent on current fashion. During the post-war reconstruction there was a clear sense that Victorian architecture was of relatively limited value and could be sacrificed. Even the high-quality Georgian buildings produced by Richard Grainger in Newcastle were not immune to this destructive urge – while the 1960s City Council retained the central area, other buildings were demolished, for example the Royal Arcade, which was replaced by an urban motorway and roundabout (Pendlebury, 1999: 427). When lamenting the losses of the 1960s, it should be remembered that not many people were shedding tears at the time. In contemporary regeneration it is clear that some buildings are seen as more worthy than others of being given a new lease of life, rather than simply being replaced. There seems to be a blanket assumption that anything built in the 1960s is automatically of limited value and fit only for the wreckers' ball.

The listing of historic buildings for preservation is the main mechanism for protecting the highest quality examples of architecture from a particular period. This is not an uncontroversial process, particularly when it comes to twentieth-century buildings. Park Hill in Sheffield was one of the first council housing megastructures in the UK designed by two young architects Jack Lynn and Ivor Smith, whose ideas can be broadly fitted into the new brutalist movement associated with Alison and Peter Smithson. Park Hill was the first built example of the deck access model in the UK – the idea that the dynamics of the working-class street could be replicated in high-rise form through everybody's front door opening on to a wide gallery which ran the length of the building. When Park Hill was listed as Grade II* in 1998 there was an outcry in Sheffield that this huge concrete 'monstrosity', which by that time had become quite a run-down estate, should be deemed worthy of preservation. Sheffield City Council found themselves reluctantly having to put together a complex partnership arrangement to find a way of preserving the estate.

Urban Splash made its name in the 1990s pioneering the redevelopment of factories and warehouses into fashionable apartments in north west England. The company has expanded its regeneration work and is now one of the most in demand developers for large, complex sites, including Fort Dunlop in north Birmingham and Royal William Yard in Plymouth. In 2006, Urban Splash were chosen to lead the redevelopment of Park Hill, creating a fashionable inner-city apartment development, reducing the number of socially rented units to one-third of the total and bringing in a variety of 'hip' new businesses. It is intriguing that the promotional material Urban Splash has produced for the redevelopment bears a striking resemblance to the kinds of photomontages that were used to promote this kind of development in the first place (Figure 6.5). The original development suffered partly because it became a sink estate for the poorest and most vulnerable tenants and partly because of severe centrally imposed restrictions on maintenance spending that left

Figure 6.5   Although Alison and Peter Smithson did not design Park Hill, their work was influential upon it. The use of photomontage in the Smithsons' design work, here for their unrealised Golden Lane scheme, bears interesting comparison with the playful computer rendering produced by Urban Splash to promote the redevelopment of this Sheffield megastructure.

the area physically neglected for many years. With a much smaller number of social housing units managed by the much better funded Manchester Methodist Housing Group, the estate has a greater chance that it may live up to the images in its promotional materials this time around.

Park Hill is not the only mid-twentieth-century building which has had its architectural and cultural value validated by a contemporary redevelopment. Denys Lasdun's 1957 Keeling House in Bethnal Green has been refashioned as exclusive apartments while Ernö Goldfinger's Trellick Tower, opened in 1972, has today become an icon of modernist chic run by a tenant management organisation. All of these buildings have, however, received listed status. There are nevertheless many good buildings of this period which, although not of the high quality required for listing, may in time be rehabilitated to fashionable status in the way that Victoriana has been. Indeed, Aidan While (2006) has discussed some of the tensions that arise where the urban renaissance agenda runs headlong into the remains of the modernist cityscape. There was a local outcry when a proposed redevelopment of Portsmouth's brutalist Tricorn shopping centre prompted consideration of its being listed. After an extensive lobbying campaign, both by the pro- and anti-redevelopment sides, the proposed listing was rejected and the Tricorn demolished. Meanwhile, Coventry's 1950s Lower Precinct, the first retail precinct of this kind, although largely retained, has received a makeover not entirely sensitive to its original post-war form.

Aside from the handful of people with an interest in this particular kind of architecture, not many lament the destruction of the built fabric of modernist culture. Again, however, it is important to reiterate that fashions and tastes change over time and those things deemed of little value culturally today, may come back into vogue at a later date. This is not an argument for blanket preservation, but rather to note that where urban regeneration talks about fostering cultural values, these are highly subjective and tend to be driven by current tastes.

Cultural anchors and signature buildings

Just as post-war architecture has fallen out of favour and can therefore be ignored or sacrificed in redevelopment, so other cultural markers are emphasised or effaced in the regeneration process, depending on how they are currently valued. A key question, therefore, is to ask whose culture is being validated through a 'cultural' regeneration process.

This question of 'whose culture' is particularly important when considering schemes where a cultural facility is being used as an anchor for a larger regeneration programme. Cultural anchors are often argued as being a useful mechanism to give developers confidence to invest in parts of cities which are run-down or unfashionable. The classic international example of using a signature building to drive regeneration activity was the 1997 Guggenheim Museum built in the declining industrial city of Bilbao in northern Spain to a design by Frank Gehry. There is a question of what the building actually says about Bilbao itself, the town and its existing cultures being hidden behind an icon in a global architecture guidebook. The building has become a tourist attraction in its own right and arguably validated a particular kind of spectacular architecture in cultural anchor buildings. A similar 'alien spacecraft' quality can be seen in the Imperial War Museum of the north which sits amid the Salford Quays redevelopment and at the Sage in Gateshead (Figure 6.6). The completion of the Sage in 2002 gave the north east an extremely high-quality concert hall, to act as home both for the Northern Symphonia and Folkworks. Folkworks, promoting traditional music and dance, in particular had been heavily involved in community and educational work since its foundation in 1988. This emphasis on community engagement continues with the educational facilities in the new building.

The eye-catching design of the Sage is clearly meant as a signature building; the bulbous metal slug was designed by the internationally renowned Foster and Partners and cost £70m. The Sage is part of a larger complex of cultural resources on the south bank of the Tyne, which includes the BALTIC, a major modern art gallery housed in a converted flour mill, and the Millennium Bridge, which links the facilities to Newcastle on the other side of the river. Ironically, perhaps, it was probably Newcastle that benefited more in economic regeneration terms in the first phase of development as developers built speculative flats on the Newcastle side of the Tyne, where Gateshead simply gained the cultural resources. Gateshead Quays Phase II is changing this. Kier Properties were picked as the preferred developer in 2006 with their proposals for a mix of retail, restaurants, cinema and,

Figure 6.6   The Sage, Gateshead, designed by Foster and Partners, provides a striking outlook to those sitting in the riverside cafés on the Newcastle side of the Tyne.

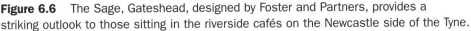

inevitably, parking along with 324 residential units to fill in the gap between the BALTIC and the Sage. There is a clear agenda here that cultural icons will lever in private investment and Gateshead Quays seems likely to be a great success economically, just as Salford Quays was before it.

Simply securing a landmark cultural iconic building is, however, no guarantee of success. The important lesson of developments such as Gateshead Quays, which hangs off the BALTIC and Sage, is that the cultural resource must itself be a success, working hard to draw visitors to an area and maintain its flow of income. There have been a number of high-profile flops, such as the National Centre for Popular Music in Sheffield and, perhaps even more dramatically, 'The Public', a community arts resource in West Bromwich (Figure 6.7). Designed by Will Alsop with an original budget of £38m, the cost went over £50m and, uniquely for a project of this kind, The Public went into receivership even before it opened. The building, a black box pierced with blob-like magenta openings, is certainly striking, but the failure to get it off the ground as an actual working resource, rather than simply a landmark building, has made it much less attractive in terms of levering in private capital. Indeed, The Public merits barely a mention on the website of the local URC RegenCo, which has a number of ongoing regeneration projects, none of which flag proximity to the building as being an asset.

Cultural quarters

As discussed in Chapter 4, the idea of 'clustering' has generated some debate in terms of its impact on local economic development. When considering cultural regeneration, it is

Figure 6.7   The Public, West Bromwich, is a lovely building to photograph, though looks somewhat less impressive when actually standing in front of it. A financial disaster, the scheme has yet to prove its value in attracting investment to this run-down part of the West Midlands.

taken as read in policy that clustering together cultural resources and the creative industries will have inherently beneficial effects. As a result, the idea of fostering cultural quarters is a key tenet of urban regeneration discourse (Montgomery, 2003). Culture is often defined quite broadly in this context, classically stretching from media companies and IT, through craft-based businesses, music, graphic design and arts organisations. Companies can range in size from the multinational IT provider to the micro-business of one person making and selling his/her own products. The idea of clustering is that by working in proximity to each other, creative people will be still more creative, with the clustering reinforcing innovation and economic development.

The classic counter-argument against cultural clustering is that it generates a process of **gentrification**. It is quite common for artists of various kinds to 'occupy' run-down parts of cities, because they are inexpensive places to live and work. The presence of this artistic culture can help re-image that part of the city and make it fashionable enough to attract other people and businesses, which eventually make the area too expensive for the artists who colonised it in the first place (Zukin, 1982). Where cultural clusters are given the official seal of approval, this gentrification process can be accelerated. Marginal and experimental artists and businesses will be squeezed out by higher rents, unless they fit into the tastes and strictures of work that attracts public subsidy. Again, therefore, notions of cultural quarters raises questions about the kinds of culture that are valued in terms of regeneration.

The Cultural Industries Quarter (CIQ) in Sheffield builds on a pre-existing cluster of cultural users. Sheffield was hit particularly hard by deindustrialisation in the late 1970s and

1980s, with the collapse of the city's steel industry. The CIQ area, running east and south from the main railway station, had been in decline even before then, having been a location for small workshops involved in the metals trade. With assistance from the City Council, a number of cultural businesses were brought into the area, including Red Tape Studios, joining existing cultural facilities such as the Leadmill live music venue which opened in 1980. The Workstation opened in 1993 with 460m^2 of space which is rented to media companies working in video, sound, design, etc. The idea is that it should act as something of a hothouse, encouraging collaboration and mutual support, providing flexible working space for the kinds of small and micro-businesses that flourish in this sector.

More high profile, but less successful was the National Centre for Popular Music which opened in 1999, housed in a dramatic signature building by Nigel Coates (Figure 6.8). Costing £15m with funds from the National Lottery, it ran into financial difficulties within seven months and closed. The buildings have now become the student union for Sheffield Hallam University. By 2000 the City Council had decided that the cultural quarter was not developing in a sufficiently coordinated fashion and produced the CIQ Action Plan, which was adopted as supplementary planning guidance. A new body, the CIQ Agency, was established to coordinate the development of the quarter, with funding secured from round 6 of the Single Regeneration Budget (SRB). The CIQ Agency has two stated aims:

1   To develop the physical landscape of Sheffield's Cultural Industries Quarter.
2   To further develop the creative business economy in the CIQ and across South Yorkshire (ciq.org.uk, accessed 11 January 2007).

In turn, the Agency's priorities can be summarised as:

- producing a coordinated approach to cultural policy and practice in south Yorkshire;
- encouraging the private sector to provide cultural workspaces rather than high-profit residential accommodation;
- delivering the Townscape Heritage Initiative, where funding has been secured from the Heritage Lottery to preserve 20 listed buildings; and
- improving public space, especially pedestrian access (derived from Dabinett, 2004: 418).

The second of these aims is very interesting as it seeks to overcome the trend whereby the fashionable nature of a district attracts residential and other land uses that price out the very sectors that made the district fashionable to begin with. An example of this is The Cube, which contains 25 live/work spaces promoted by the regional development agency Yorkshire Forward as a demonstration project. This development is explicitly promoted as giving entrepreneurs an opportunity to deeply embed themselves in the networking opportunities offered by the density of cultural businesses in the CIQ.

By delineating this area as a specific 'quarter', the CIQ gives this area an identity around which to frame regeneration within the context of the broader regeneration of the city. The project to produce an integrated transport hub at the adjacent train and bus stations, which was coordinated by the local URC, for example, keys into the neighbouring CIQ as one of a number of development quarters being opened up. The CIQ is also adjacent to the Sheffield Digital Campus, which is conceived as a city-centre business park for small and medium-sized firms working in the ICT sector.

Figure 6.8   Originally the National Centre for Popular Music, Sheffield, this collection of futuristic curling stones is now the students' union for Sheffield Hallam University and sits at the heart of the city's Cultural Industries Quarter.

The idea of promoting 'cultural industries' is clearly a key strategy among regional development agencies and this returns us to the question about the kinds of cultures being promoted through these processes. New media, ICT, television and film appear to be the holy grail in terms of economic regeneration through the cultural industries, none of which necessarily reflect or develop the unique cultures of local areas. Where culture is only seen as a means to an economic end, there is a danger that the somewhat ephemeral qualities of creativity which cultural quarters seek to foster will be lost because of an external conception of what are the 'right' kinds of culture to encourage regeneration.

Sub-cultural quarters

There is a strange tension in the notion of a cultural quarter, because the idea of quarters suggests a space given over to a particular *sub*-culture, one part of the rich tapestry of cultural life in a city. Traditionally, quarters or districts were associated with ethnic groups – Chinatown, the Irish quarter, and so on. The idea of a *general* cultural quarter is therefore a little peculiar, but perhaps reinforces the idea that culture in this context means mainstream, economically productive 'creative' industries.

Ethnically-defined clusters have been the target of regeneration activity, but it is perhaps gay quarters that have become the most iconic of the sub-cultural clusters in contemporary urban regeneration. The sheer size of cities has tended to make them more 'gay friendly' and much more tolerant of diversity. Larger populations have also enabled critical mass to develop in certain areas, with specifically gay shops and services catering to the needs of a population clustering in part to offset the negative consequences of homophobia.

Given that gay couples are less likely to raise children, this produces a tendency towards reducing lifetime housing costs, freeing up resources for consumption. This in turn makes the 'city living' lifestyle more attractive, with proximity to specialist services becoming more important than issues of child safety, open space and good schools that make suburban areas attractive. Alan Collins (2004) has examined the formation of gay villages and their role in urban regeneration. While many areas may have some of the precursor conditions for establishing a gay village, such as a large diverse population and a low rent district where more specialist businesses can get established, Collins argues that it takes a kind of historical accident to transform these precursors into fully-fledged clusters. This might be, for example, a micro-economic decision such as opening a gay or gay-friendly bar in a particular district, which leads to cumulative processes attracting similar business where a market need is demonstrated, which in turn establishes the area as one where gay households want to locate.

While gay villages such as Newcastle's 'pink triangle' or London's Soho may begin to develop in this rather organic way, the economic potential of the pink pound has meant that policy-makers and developers have taken an interest in the gay market. Canal Street in Manchester is the archetype of a previously marginal space which has been heavily invested in and is now promoted by the City Council as one of the city's major tourism/leisure attractions. In the late 1980s, the Canal Street area had nowhere near its current visibility or coherence, consisting of just three traditional gay bars with proximity to two major cottages – locations for casual, semi-public sexual activity among gay men. This last point is important because it indicates that certain sub-cultural phenomena, with which a local authority would not want to be associated, can help to incubate the more politically acceptable manifestations of that sub-culture. The opening of Manto in the early 1990s gave Canal Street an architecturally designed, highly *visible* gay bar and started a trend for major investment. Pedestrianisation of the area followed, complimenting a street café culture and expensive loft apartment development occurring by the mid- to late 1990s (Binnie and Skeggs, 2004). A previously marginal area thus became very fashionable and developed rapidly on the back of its association with a particular sub-culture.

Marketing Manchester, a Public Private Partnership (PPP), developed a specific marketing campaign in 1999 based on the city's gay friendliness. This campaign was helped significantly by a TV series broadcast the same year, *Queer as Folk*, which was filmed in Manchester. The image of Manchester as a gay idyll made the series a cult hit in the United States and helped in the city's broader international re-imaging process (Hughes, 2003). The way that the Canal Street quarter has developed has not been uncontroversial, however. The area was one of the first in the city to develop the model of more continental-style late night café culture and, as such, was rather fashionable for a time. It also became popular with women as a location where hedonistic impulses could be indulged without the threat of an aggressive heterosexual male presence. Unfortunately, this very fact attracted straight men into the area, resulting in somewhat of a backlash, with gay-only entry policies and tension within the community with the feeling that straight people were coming to voyeuristically observe gay behaviour as if visiting a zoo. Attempts by mainstream cultures to co-opt sub-cultural spaces have the potential, therefore, to produce significant tensions and can indeed threaten the very sub-cultural qualities that had made the area attractive in the first place.

Liverpool, European Capital of Culture 2008

The European City (later 'Capital') of Culture (ECoC) scheme began in 1985 and in 1990 Glasgow was the first UK city to be awarded the designation. While this raised some eyebrows, particularly among some cynical (particularly English) commentators, ECoC designation provided an opportunity for Glasgow to shift focus away from its image of being a deprived post-industrial city and to remind people of its rich cultural heritage. Among policy-makers, the Glasgow experience was deemed a huge success, although as discussed in Chapter 4, it was not uncontroversial and there are those who continue to argue that it did not really have that much of an effect in terms of social regeneration (Mooney, 2004). Regardless of the arguments on the success or otherwise of economic restructuring, in terms of city re-imaging, Glasgow's reputation has been transformed post-1990 and it is little surprise that when it was again the UK's turn to host ECoC, there was fierce competition between cities for the 2008 designation. When the result was announced in 2003, it was Liverpool that came out on top.

Like Glasgow, Liverpool today has a significant disparity between a revitalised, exciting and well-resourced city-centre regeneration catalysed by notions of 'culture' and truly diabolical poverty in its inner urban and suburban social housing estates. One of the stories that was told in the aftermath of Liverpool's victory in the ECoC competition, was that the significant public involvement in the bid process was the factor that tipped the scales. It has since been characterised as the 'people's bid'. Jones and Wilkes-Heeg (2004) argue that perhaps a more important underlying factor was the belief that Liverpool was the city most likely to 'do a Glasgow' by using the Capital of Culture label as the lynchpin of a broader regeneration process. Rival cities, such as Oxford, Birmingham and Bristol, though hardly free from poverty, arguably would not benefit as greatly from the additional impetus generated by the Capital of Culture label in their ongoing regeneration programmes.

In the run up to 2008 there were claims that the Capital of Culture would bring 12,000 new jobs to Liverpool while doubling visitor numbers to 38 million per annum and generating £2bn of extra spending in the local economy. The bid team's consultants, ERM Economics, had in fact painted a much less rosy picture. They estimated approximately 720,000 extra visitors per year, rather than 19 million. Indeed, though the report did talk about an extra 13,200 new jobs, this was a projection of general growth in the creative industry sector – jobs directly created by the ECoC were estimated at less than 1,400 (Jones and Wilks-Heeg, 2004). One can, however, understand the boosterist claims given that Liverpool has been experiencing chronic unemployment and population loss for decades. No doubt consultants and academics will spend many years post-2008 analysing the regenerative success or otherwise of Liverpool's ECoC year.

In many ways, ECoC can be seen as the latest in a line of explicitly culture-driven regeneration programmes in Liverpool that began with the Albert Dock. The Ropewalks area of the city, just to the south of the central business district (CBD), was investigated as a possible area for a regeneration scheme through the 1990s and eventually the Ropewalks Partnership was established to carry forward a plan produced by the Building Design Partnership. With European Union Objective 1 funding secured through the Government Office for the North West as well as resources from the North West Regional Development Agency, a major programme of new construction, refurbishment of historic buildings and improvements to the public realm were undertaken to cement Ropewalks as a cultural quarter. The anchor building for this programme was the FACT Centre (Foundation for

Art and Creative Technology). Established in 1988 as 'Moviola', FACT opened a new building in 2003 boasting exhibition and work spaces, training courses and a state-of-the-art
cinema with 70mm projection and THX sound (Figure 6.9). Ropewalks was one of the
developments flagged in the ECoC bid to indicate that 'culture' in Liverpool was not merely
trading on the city's historic legacies.

There is some irony in that one of Liverpool's more interesting, grass-roots cultural institutions, Quiggins, has come out rather badly from the redevelopment of the city centre.
Founded in the mid-1980s, Quiggins occupied a large, somewhat ramshackle building on
School Lane, a then low-rent area just behind the city's main shopping axis. Home to a variety of 'alternative' traders, Quiggins served as a central city business incubator for small
craft designers and other micro-businesses. Quiggins was approximately 200 metres
outside the boundary of the Ropewalks cultural cluster, but well within the Paradise Street
Development Area and was threatened with a compulsory purchase order in 2004 which
persuaded its owners to sell up. Many of the traders moved to a new home, 'Grand Central',
housed in the old Methodist Central Hall, although this is somewhat further out from the
city core. While the façade of the Quiggins building will be retained in the new development, it will simply form part of the very large new shopping complex which is being built.
Quiggins, effectively, was the wrong kind of culture in the wrong place.

The Quiggins issue was a very minor distraction in the rapid pace of change in central
Liverpool. Liverpool Vision was set up in 1999 as the first URC in the UK and has helped
to bring together a number of schemes in the city, including the Paradise Street redevelopment, which is branded as part of Liverpool One (see Chapter 4). Liverpool Vision was
a key partner in the last major piece of the ECoC puzzle in the city, the Kings Waterfront
development. At the heart of this is ACC Liverpool, an arena and conference centre. This
£146m development was designed to play a key part in the ECoC celebrations and consists of the 10,600-seat Liverpool Echo Arena (named for a local newspaper) as well as
7,000m^2 of exhibition space (Edwards, 2007). Again, Objective 1 funding was a major
source of finance for this, although the ACC scheme has levered in a significant amount
of private capital for apartments, a hotel and other developments anchored by this key
cultural resource.

Sports-led regeneration

ACC Liverpool gives the city a major venue for hosting indoor sporting events and reflects
a trend which considers sport not as a cultural phenomenon in its own right, but as a major
potential source of urban regeneration activity. When discussing the potential benefits of
sport, the examples of Barcelona and Sydney are often cited. A decade on from the
Barcelona Olympics, hotel capacity and tourist numbers in the city were still virtually double the pre-Games figures. Yet, as Gratton et al. (2005) have argued, neither Barcelona or
Sydney had the problems of declining industrial economies that typify many of the areas
targeted in the UK for sports-led regeneration. Indeed, both cities were already major
tourist destinations that did not need to be drastically re-imaged and so the validity of the
comparison is somewhat suspect.

One of the first cities in the UK to explicitly see a sporting event as an opportunity to undertake regeneration activity was Sheffield, which played host to the 1991 World Student Games.
With the collapse in steel making in the Don Valley area, by the mid-1980s key members of

Figure 6.9   FACT, Liverpool, acts as the key anchor in the city's Ropewalks development and has been fitted sensitively into the existing narrow streetscape.

the City Council's ruling Labour group became committed to the idea that sports and leisure activities could be one way of rebuilding the city's shattered economy (Henry and Paramio-Salcines, 1999). The successful bid to host the World Student Games was seen as the jewel of this strategy, though it was not uncontroversial in the city. With caps on local government spending imposed in 1990, some Labour councillors in Sheffield were unhappy at resources being diverted to this new sports-led strategy at the expense of conventional social programmes, with most of the £147m capital costs being met by the Council. Indeed, although the city has ended up with world-class sports facilities clustered in the Don Valley, this has been at the expense of some community-level facilities, particularly for swimming, which were closed to help pay for the centralised resources (Henry and Dulac, 2001).

The Sheffield experience was somewhat problematic in regeneration terms, partly because the focus was on delivering the sporting event rather than the broader legacy. In this context, the Manchester Commonwealth Games of 2002 provides a more interesting example. The spend on sporting infrastructure in Manchester was around £200m, making it the largest ever investment in sports hosting in the UK prior to the London Olympics. In addition, however, a further £470m was spent on non-sports infrastructure as part of a major redevelopment programme in east Manchester.

The 2002 Commonwealth Games set the standard for UK sports-led regeneration by attempting to ensure that there should be clear, long-term benefits to the economy of east Manchester once the event was over. In order to achieve this, the Commonwealth Games Opportunities and Legacy Partnership Board was set up in 1999 to manage the legacy of the programme. After the Games were complete, the sporting facilities were branded as Sportcity – the stadium became the main ground for Manchester City football club while the other sporting venues have been used to establish the English Institute of Sport, providing elite training facilities and generating an ongoing revenue stream in the area. This careful planning has avoided the problem of 'white elephant' facilities with no clear post-Games purpose – a lesson that the Olympics in Athens could well have learned.

In addition to securing a use for the Sportcity facilities, the New East Manchester URC was established in 1999 to coordinate the regeneration of east Manchester more generally, an area covering some 1,100 hectares. The URC's initial aims were to:

- double the population to 60,000 over 10–15 years;
- build up to 12,500 new homes offering a range of tenure and type;
- improve 7,000 existing homes;
- create a 160-hectare business park;
- create a new town centre with 11,000 m² of retail provision;
- produce an integrated public transport system;
- create a new regional park system;
- bring educational attainment above the city average (New East Manchester, 2001).

In order to realise these broader aims, New East Manchester URC attracted £25m from SRB round 5, £52m from New Deal for Communities and around £3m from SureStart. In addition, the URC has attracted private capital to bring new facilities to the area, including a regional-sized Asda supermarket. In 1999 the housing market in the area had some severe structural problems and, in some districts, the market had completely collapsed. The value of the Sportcity brand as a marker of redevelopment in the area helped the URC to bring in developers Countryside Properties, who produced a residential masterplan and have worked on a series of new housing schemes. Clearly, east Manchester is still an area suffering from multiple deprivation, but there can be no doubt that the sporting event and associated infrastructure have been used as a catalyst for a more general economic and social regeneration involving a mix of public and private resources.

One of the interesting differences between the UK and United States is that sporting teams in the UK tend to be far less 'footloose'. Middle-ranking cities in the USA frequently compete to attract sports team franchises to their cities, offering tax breaks, large new stadium complexes and other incentives. The economic benefit to the community of this kind of investment is somewhat dubious, particularly as there is little to stop teams moving once again if another city offers an even better deal (Crompton, 2001). The UK has therefore by and large escaped the phenomenon of cash-strapped local authorities giving subsidies to multi-million pound sports businesses – the move of the MK Dons football club from London to Milton Keynes being very much the exception rather than the rule. Indeed, there have been some examples of individual sports teams engaging in regeneration activity, such as when Arsenal Football Club moved to a new stadium. The club teamed up with the local authority and other partners to undertake a series of new developments. Newlon Housing, a social housing provider, was brought in to help meet a target for 25% of the 2,500 new homes built in the developments to be affordable – attempting to tackle an acute shortage of affordable homes in the club's neighbourhood. Some of the developments took place on the site of the old stadium, part of which was listed for preservation and has since been converted into new homes. The old pitch itself has become a memorial garden, respecting the memory of the fans whose ashes had been scattered there. Business units for small and medium-sized enterprises have been created in the area and the Arsenal Regeneration Team, which includes representatives from the club, have also promoted other community development projects.

The Arsenal regeneration has been relatively modest, with the stadium redevelopment project giving impetus to regeneration targeted at the local community. This stands in stark contrast to the works associated with the 2012 London Olympics, with the then Culture Secretary Tessa Jowell admitting in 2007 that the estimated cost of the Games had increased from £2.4bn to £9.3bn – although the £1.7bn directly allocated for regeneration remained constant. The Games have proved immensely controversial, not least because some of the projects that have been folded into the legacy outputs were happening anyway. On the transport front, phase two of the Channel Tunnel Rail Link, although construction started in 2001, has been rebranded the 'Javelin train', while prior to winning the bid Transport for London had explicitly stated that the East London Line extension would happen regardless of whether the Olympics came to London or not (Gilligan, 2007). A major shopping development with associated residential accommodation, Stratford City in Newham, was at the planning application stage in 2003 – again, prior to the bid proposal. At the same time the escalating costs meant cuts to other agencies, including Sport England, which meant reduced funding for programmes to increase sporting participation at the community level – a key target in social regeneration (Bond, 2007).

The Olympics meant that a great amount of land in the Lower Lea Valley area was bought for redevelopment rather more rapidly than might otherwise have been the case. This has been facilitated by changes to the laws on compulsory purchase, which were made in advance of the successful bid, in part, it was argued, because if the bid was successful there would otherwise be delays to the project as the laws were rewritten. Compulsory purchase is a very time-consuming and controversial process, meaning that local authorities and those development agencies granted these powers generally prefer to negotiate with landowners, with compulsory purchase being seen as a last resort. A number of extremely positive reforms were contained within the Planning and Compulsory Purchase Act 2004, including placing a statutory requirement on planners to 'contribut[e] to the achievement of sustainable development' (section 39, 2). There was also, however, an amendment to section 226 of the Town and Country Planning Act 1990, to include a general 'well-being' clause. Compulsory purchase can be justified under the new Act if the purchase can be demonstrated to improve one or all of the economic, social or environmental 'well-being' of the area (section 99, 3). The rather hazy idea of 'economic well-being' can be seen as giving local authorities more leeway when it comes to securing compulsory purchase. In the Olympic case, compulsory purchase powers were granted to the London Development Agency to acquire the 306 hectare site. The vast majority of residents and businesses were moved through negotiation, but a hardcore of individuals resisted. The last of these, two groups of Gypsies and Irish Travellers who had been long-term residents of the area, lost their appeal against the compulsory purchase order in May 2007 (Williams, 2007). The changes to compulsory purchase legislation, however, have wider impacts than a politically important national project happening in London, and apply across England and Wales. While compulsory purchase remains a time-consuming process, the reformed legislation makes the threat of compulsory purchase a much more potent bargaining counter when negotiating with reluctant landowners.

Key points

- There is a tension between culture as an aesthetic phenomenon, such as buildings and artistic performances, and culture as a way of life.
- The transition to a post-industrial economy has opened up major redevelopment opportunities on former industrial sites, with dockland redevelopment having become almost a cliché of urban regeneration.
- The re-use of historical buildings can give a redevelopment significant character but it is also expensive and subject to the whims of fashion in terms of which kinds of architecture are valued at any given time.
- An expensive signature building can be an embarrassment, rather than an anchor to new development if the resource it houses is not successful.
- Cultural quarters have become quite fashionable among those promoting regional economic development attempting to foster creative industries, but there is a risk that non-officially sanctioned cultural expressions will be forced out, through a combination of gentrification and mainstreaming.
- Sports events and facilities have been successfully used to anchor urban regeneration schemes, but it is crucial to consider legacy uses.

Conclusion

There is a tension in this chapter, and in regeneration more generally, between 'culture' as something which can be exploited for economic development and something which is of value in its own right as part of a diverse and healthy society. This, in part, comes because the word 'culture' has such diverse meanings, that a cultural regeneration can contain contradictory claims for how it promotes and uses culture.

When thinking about culture as representing a way of life for individuals and communities, regeneration does not always treat cultures particularly well. The 'wrong' kinds of cultural expressions located in an area which has been earmarked for redevelopment can find themselves being squeezed out, even if they were the source of the area's attractiveness for development in the first place. As was demonstrated by the example of gay villages, sub-cultural characteristics can become threatened as elements of that sub-culture start to be co-opted by the mainstream.

Notions of culture are also frequently wrapped up into the notion of the knowledge economy, with cities competing to attract workers and businesses in IT, the media, arts and similar sectors. Partly, this is manifested through attempts to improve the public realm and attract flagship cultural functions to make the city more appealing to the so-called 'creative class', as well as tourists. The other key manifestation of this is the creation of cultural quarters, attempting to cluster business working in the creative sector in order to produce a hothouse of talent built on face-to-face networking. Here culture is specifically defined as that which will be economically productive within the post-industrial model. As these quarters mature, more financially marginal arts and enterprises will be driven out, except where an external body, such as a local authority or regional development agency, deems them worthy of subsidy.

Aesthetics plays an important role in ideas of culture. Attractive, flagship buildings produced by internationally renowned architects have become a feature of many regeneration projects, with the hope that they will create an image of the district which is innovative, exciting and worthy of investment, acting as an anchor to further commercial development. In some regards, however, the buildings are only as successful as the uses to which they are put, and high-profile flops have left some areas with embarrassing and very expensive white elephants for which new uses have to be found. The same is true of historic buildings which are put to new uses, with the added complexity that an insensitive conversion can destroy some of the coherence of that building as a manifestation of the culture that originally built it.

Another problem with aesthetics is that the attractiveness of a development is entirely subjective, with changing tastes making it very difficult to audit the 'quality' of an aesthetic design. Some attempts have been made to regulate both the functioning of urban design (for instance, asking if the road layout encourages car use) and also its aesthetics. Design coding is one tool, developed by the new urbanists, for producing more coherence between the architecture and the character of an area at the district scale. Although design coding is now in the policy mainstream, it is no guarantee of a 'good' aesthetic product.

The complexities and contradictions of cultural regeneration aside, culture is in the mainstream of policy **discourse** and will likely play an ever more important role. Undoubtedly, some cultures – both ways of life and aesthetic products – will be more valued than others by decision-makers undertaking regeneration projects, and one should question the kinds of culture that end up being emphasised. As a final note it is worth reflecting that the authors of this book are both middle-class, professional, white men in their thirties with no children, which gives us a very specific, subjective position on the kinds of culture that we value and discuss.

Further reading

A great deal has been written on urban design, culture and regeneration. Of the many publications produced by the Commission for Architecture and the Built Environment, *By Design* (CABE, 2000) gives a good introduction to the kinds of urban design principles which are seen as best practice, while Bell and Jayne (2003) give one of the best critiques of design-led regeneration as a whole. Marshall (2001) gives a good introduction to the whole notion of how post-industrial waterfronts have been redeveloped. It is also worth revisiting Florida (2002) to understand some of the economic ideas which underpin the whole notion of cultural clustering and the creative city. The special issue of the journal *Urban Studies* (2005) on culture-led urban regeneration contains a number of high-quality academic papers on the topic, generally with a UK focus. Gratton has written extensively on the role of sport in society and has edited a collection of essays with Ian Henry examining sport and regeneration (Gratton and Henry, 2001). Although not limited to UK examples, it provides excellent case study material and analysis.

Bell, D. and Jayne, M. (2003) 'Design-led urban regeneration: a critical perspective', *Local Economy*, 18: 121–34.

CABE (2000) *By Design. Urban Design in the Planning System: Towards Better Practice* (DETR, London).

Florida, R. (2002) *The Rise of the Creative Class and How It's Transforming Work, Leisure, Community And Everyday Life* (Basic Books, New York).

Gratton, C. and Henry, I. (eds) (2001) *Sport in the City: The Role of Sport in Economic and Social Regeneration* (Routledge, London).

Marshall, R. (ed.) (2001) *Waterfronts in Post-industrial Cities* (Spon Press, London).

Urban Studies (2005) 'Special Issue: The Rise and Rise of Culture-led Urban Regeneration', *Urban Studies*, 42(5–6).

7 Regeneration beyond the City Centre

Overview

Policy and academic focus has hitherto concentrated on the city centre. This chapter charts the extension of the urban regeneration agenda beyond central cities, to the suburbs and beyond.

- *Suburban regeneration*: Can the central city regeneration model be transposed onto the suburbs? Explores case studies of Solihull's Chelmsley Wood estates and Dalmarnock in Glasgow.
- *Ex-urban*: examines regeneration beyond the suburbs with case studies of Port Marine in Portishead, Waterfront Edinburgh and Lightmoor in Telford.
- *Scaling up: mega regeneration in the Thames Gateway*: looks at the project for the enormous new development stretching to the east of London.

Introduction

The definition of what comprises a 'suburb' can be somewhat fuzzy. In north America, the word is often used to refer to those outer parts of a built-up area which are beyond the political administration of central city authorities. In the UK, 'suburb' has become a catch-all phrase referring in general terms to the outer city or that which is beyond the city core, though normally within the same political territory (Whitehand and Carr, 2001). Thus defined, suburbs are home to around 86% of the UK's population and as such are exceedingly heterogeneous. Surprisingly, however, there has been relatively little academic work to date which deals specifically with the regeneration of suburban areas. In more recent years there has been some attempt to address this problems, with two major research teams having been set up with a specifically suburban focus: the Centre for Suburban

Studies at Kingston University and the SOLUTIONS (Sustainability Of Land Use and Transport In Outer NeighbourhoodS) project based at the Martin Centre, University of Cambridge. SOLUTIONS has focused on a modelling approach, particularly of different transport planning scenarios, where the Centre for Suburban Studies has taken a more social and cultural approach to the suburb and its development. The broader issues of regeneration in the suburbs and the extent to which they pose a different challenge to the central city has, however, received fairly limited attention.

In the UK, historically, the idea of suburbia has had some rather negative connotations. Prior to the development of rapid transportation, the suburb, the area beyond the town, was where the poor lived. With changes in transport technology during the nineteenth century, suburbs became the place where the well-to-do could escape from the noise and smell of the central city. As cities expanded, so negative images began to be associated with them, suburbs becoming characterised as a sprawling cancer of bricks spreading out across the British landscape. In the 1920s and 1930s there was a further wave of expansion following the new arterial roads being built to service the motor car and omnibus. The spreading town was held in opposition to an image of an unspoiled rural idyll, an image which still affects a great deal of our understandings of the tension between town and country in the UK. Suburban expansion triggered the widespread use of green belts, formalised in a government circular of 1955, to restrict further outward growth, although suburban development continued post-war, filling in the gaps between the edge of towns and their green belts. It must be emphasised, however, that both during the inter-war and post-war building booms, it was not just large houses for the middle classes that were being built, but also very large estates of council housing, which had a distinct demographic profile and pose distinct challenges for urban regeneration today.

The way that towns have developed historically means that the suburbs form an exceedingly heterogeneous group of different land uses, house types, demographic groups and environmental qualities. One of the most useful studies of suburbs and regeneration was produced for the Joseph Rowntree Foundation (Gwilliam et al., 1999) and attempted to give some critical form to this heterogeneity by producing a typology:

- historic inner suburb;
- planned suburb;
- social housing suburb;
- suburban town;
- public transport suburb; and
- car suburb.

This typology is far from perfect: both wealthy gentrified areas and poorly maintained districts housing large numbers of socially deprived people could fall under the category of 'historic inner suburb' and yet require very different degrees of regeneration intervention. The typology is useful, however, in breaking down the idea that suburbia is a monolithic category, with suburbs home solely to the white middle classes.

Hampshire County Council leads the In Suburbia Partnership which operates between a series of county councils and the Civic Trust to produce guidance on more sustainable approaches to suburbia. This Partnership has adopted the typology from the Joseph Rowntree Foundation report and used it to inform their discussions about the diversity of the suburban experience. Examining the specific needs of suburban areas, the Partnership has outlined a set of principles for ensuring sustainability and a high quality of life for residents:

- an appropriate and stable context;
- continuous improvements in environmental sustainability;
- good-quality, affordable housing, with more choice in tenure and type of house for people of all ages and social groups;
- choice in mode of transport, so that walking, cycling and public transport become more viable;
- access to good-quality local services and facilities;
- a community hub or heart;
- a diverse local economy with jobs for local people; and
- social inclusion and community safety (In Suburbia Partnership, 2005: 5).

There is a clear stress here on transport infrastructure, in particular transport choices beyond the private car. Similarly, there is an emphasis on local employment to reduce the dependence on commuting outside the area. There are clear commonalities here with the north American idea of 'smart growth'. The context for development in the United States is somewhat different, with uncontrolled outward sprawl of cities still a major problem and the principle of state intervention to better regulate development not as well established as in the UK. Nonetheless, the emphasis on community, mixed uses, compact building design, walkability and public transport options are part of mainstream policy discourse in the UK. A new wave of housebuilding on the edges of towns and cities, particularly in England, makes these lessons ever more important.

This model of local provision combined with good connectivity fits closely with the 'suburban town' type identified by the Joseph Rowntree Foundation report (Gwilliam et al., 1999). As towns and cities have grown outward over time, smaller settlements on the edge of the urban area tend to be absorbed. London is sometimes described as a city of villages because of the way that historically distinct settlements have retained their identities since being swallowed up by the expanding conurbation. Places like Camden, Greenwich and Kew were all urban settlements in their own right before they became buried in London's middle suburbs. These areas have retained their own identities and retain many of their independent functions, acting as towns within towns, with shops, services and sources of employment as well as good public transport links. At the same time, because these areas are still on a relatively small scale, they can function as walkable settlements. In a sense, therefore, the holy grail of urban regeneration – the small-scale, sustainable mixed development – is already in place. This is not only true of London; in all of the major cities it is easy to identify local sub-centres that have developed from historically distinct settlements. These areas provide good models on which suburban regeneration can build.

The notion of a polynucleated settlement is making its way into planning discourse, whereby multiple centres within the urban area are each self-contained to a degree. The provision of local services is critical, however, because suburban areas lack the competition for service provision that, theoretically at least, helps maintain quality and price in central areas. The major challenge for the polynucleated city of suburban towns is the extent to which big out-of-town shopping centres are out-competing them while at the same time increasing the dependence on the private car. There has been a reaction in the UK against the planning policy of the late 1980s which encouraged out-of-town development and the value of suburban centres has been recognised. Unfortunately, however, there is a tension here with the current direction of planning policy as laid out in the Barker report on land use planning (Barker, 2006). This report, with a number of caveats, has called for the selective release of development land on the edge of existing areas to meet demands for growth. There are sound economic reasons for this and, indeed, there is some environmental justification to reduce commutes for people who otherwise choose to live beyond the green belt. The risk is that unfettered, market-driven development on the fringes of urban areas will prompt a return to car-dependent models which out-compete in-town provision.

Suburbia does possess a particular set of associations in the *English* psyche. Where continental Europe, and, indeed, Scotland, is more comfortable with the notion of apartment dwelling throughout the whole lifecourse, in England there is a sense that families with children should be based in a house with a garden. As a result, the surburban 'semi' remains a general aspiration. This has meant that although the city-living model has taken off over the past ten years, it is as a distinct 'young' phase of the lifecourse, with city-centre residents twice as likely to be single as the national average and two-thirds of whom are aged 18–34 compared to the national average of a quarter (Nathan and Unsworth, 2006). This has been a factor encouraging the relatively rapid turnover of central city populations, as against the comparative stability of communities in many suburban areas.

Max Nathan and Rachel Unsworth (2006) have identified the historic inner suburbs as representing a key opportunity to smooth the churn of city-centre residents moving out of the inner city as they get older and plan families. This is particularly an opportunity within northern cities where some inner suburbs have suffered from demand problems of the kind which are being targeted by the Housing Market Renewal Pathfinders (see Chapter 2). Some developers have, however, started to apply the successful city-centre model to these inner suburbs, building high-density blocks of studio and one-bedroom flats. There are questions about how sustainable this is, given that these districts are often slightly too far from the centre to truly give walkable access to central resources and therefore lose one of the key selling points of the city-centre experience. Similarly, the larger properties, typically Victorian terraces, that characterise these areas can act as much needed, comparatively inexpensive, family accommodation in the inner city. Demolition of such properties to be replaced with smaller flats thus damages the potential for demographic mixing within these areas and this has been a key critique of the Housing Market Renewal Pathfinders. At the same time, there are broader issues to consider, such as the quality of local schools, which is often a major determining factor for location among middle-class families. Regeneration of these areas, which seeks to attract a proportion of wealthier residents, needs to take account of these social needs as well as market potential.

What is highlighted in this discussion is that the highly profitable central-city model of building small apartments is not necessarily appropriate to suburban areas. This is not to say that developers will not be able to make money selling these units, but rather that this will not necessarily help tackle some of the broader social needs that regeneration seeks to address. There is a question, therefore, of whether the model of regeneration in the UK is overly **metrocentric**, where a style of redevelopment that works well in the central cities is being inappropriately shoehorned into suburban areas.

The remainder of this chapter explores this question through a series of case studies at three different scales: suburban developments within the city boundaries; 'ex-urban' developments which are detached from the city but have a clear relationship to it; and mega-regeneration, where an entire region is being reconfigured.

Suburban regeneration

The social housing suburb is one of the categories identified by the Joseph Rowntree Foundation. These areas are the main targets for suburban regeneration, not least because established, economically successful middle-class areas are not priorities for major interventions. Social housing suburbs themselves come in distinctive types. Those estates built before or just after the Second World War generally comprise large three-bedroom houses, many of which are now in private ownership after they were sold under the right-to-buy legislation of the early 1980s. While these estates are not problem-free, many are characterised by relatively stable and somewhat demographically mixed communities. The estates built in the 1950s and 1960s have tended to fair less well, given that these are even further out from central city services and employment, and were often built with experimental techniques and unpopular building types, such as the maisonette[1] and tower block (Jones, 2005). Because of their isolated location and the physical defects of the dwellings, these estates became difficult-to-let, meaning that only the poorest and most vulnerable would accept the offer of accommodation on them. As a result, many larger city councils have been left with sink estates on their peripheries, some of which have been transferred into the hands of housing associations under the Large Scale Voluntary Transfer (LSVT) mechanism (see Chapter 2). It is these social housing suburbs which pose the greatest regeneration challenges in the urban periphery.

Glasgow 2014

Glasgow's east end contains areas representing some of the most acute social deprivation in the European Union. Estates like Drumchapel, Easterhouse and Red Road became notorious for chronic unemployment, ill health, drug addiction, crime and poverty. Many of these areas received central government funding in the 1980s and 1990s for large-scale demolition and refurbishment programmes, but problems persisted. In 2002, Glasgow City Council successfully transferred ownership of its housing stock to the Glasgow Housing Association (GHA), which operates through 62 local housing organisations. Stock transfer unlocked a number of funding routes, including private sector finance, and the degree of local control offers the possibility of better targeted regeneration activity involving the local community, which has representatives on the boards of the local housing organisations (Daly et al., 2005).

The Dalmarnock area of the city has its fair share of the social and economic problems that plague the east end. A bid for the 2014 Commonwealth Games was the catalyst for a series of long-considered major infrastructure investments in this part of the city, levering in a large amount of funding from the Scottish Executive. The controversial M74 extension as well as major new arterial link roads are being built to make the area more connected to the rest of the Glasgow. The demolition of a number of high-rise and other housing blocks owned by the GHA provides an opportunity to build a major new suburban settlement with over 1,000 new homes. Close to the Celtic Park football stadium, this development sits on the northern bank of the River Clyde about 5km from the city centre. The bid envisioned this area as the Athlete's Village, to become a **mixed development** of privately-owned and GHA-rented properties after the Games is complete.

The advantage that Scotland has over England is the greater willingness among Scots to consider apartments as an acceptable housing type for families with children. This gives the designers looking at Dalmarnock more freedom to use the kinds of apartment development familiar from city-centre regenerations, although in Scotland, as in England, central developments have still tended towards producing relatively small units for households without children. Where city-centre developments can rely on proximity to central services, suburban projects need to think carefully about the kinds of facilities that are available in the local area. The Dalmarnock redevelopment will need to consider whether local shops, schools and other services in the local area are of sufficiently high quality to appeal to the middle-class families which it hopes to attract to buy the private housing. If these facilities are not up to standard, the development runs the risk of simply becoming a car-based commuter settlement, with all the resultant implications for sustainability. At this relatively early stage in the development proposals, it is unclear quite what the mix of dwelling units will be and what kinds of additional facilities and services will be brought into the area to cater for the new population. Nonetheless, Dalmarnock provides an interesting example of how an event-based regeneration can be used to target a social housing suburb.

North Solihull

The plans for north Solihull are rather more advanced than those for Dalmarnock. Solihull is a somewhat divided town. Most of Solihull is quite wealthy, its suburbs falling into the cliché of leafy, middle-class enclaves. The northern part of Solihull is, however, quite different. Following the reorganisation of local authority boundaries in the mid-1970s, Solihull was given control over a very large suburban housing estate which had been built by Birmingham City Council in the late 1960s. The estate was built very quickly – at the time the local authority proudly boasted that it was the size of a Mark I new town, but built in just five years. Unlike the new towns, however, the careful mix of shops, services, sources of employment and demographics was somewhat lacking. Although, quite innovatively for the time, a proportion of the houses were built for sale, the majority were for local authority tenants. The estate was served by a number of small shopping centres and retained a somewhat isolated feel, fenced in to the north and east by the M6 motorway and to the south by Birmingham International Airport. While parts of this vast estate have faired well, others have experienced the classic symptoms of areas with large concentrations of socially deprived residents.

In order to tackle the problems of this area, Solihull Metropolitan Borough Council have established the North Solihull Regeneration Partnership. A 15-year regeneration project is being undertaken, covering an area containing more than 15,000 households. The local authority have taken the lead on this, but have brought in Bellway Homes, InPartnership Ltd. and the Whitefriars Housing Group as part of a Public Private Partnership (PPP) to undertake the redevelopment. The stated aims of the project are quite interesting:

- changing almost 40,000 people's lives for the better;
- £1.8bn public and private investment over the next 15 years;
- 8,500 new modern homes;
- new, state-of-the-art primary schools; and
- vibrant village centres delivering key services (Solihull MBC, 2007).

The fact that the primary aim of this regeneration is not creating a whole new urban environment but actually improving people's lives, is an unusual emphasis in the rhetoric surrounding these kinds of project. The fact that some 8,500 new homes are planned in what is already quite a densely built-up residential area indicates that a large number of demolitions are anticipated, particularly targeting some of the estate's 34 high-rise blocks. Given that many of the new homes being built will be for private sale, however, indicates that there is a clear mission to change the demographic mix of the area, reducing the concentration of tenants in socially rented accommodation.

The estate was built on a greenfield site and so did not absorb older settlements which might have been used to form a suburban town-type community hub. The estate was originally broken down into a series of sub-settlements, nominally with their own identity, but these were quite large and not particularly distinctive. The solution adopted by the North Solihull Regeneration Partnership is to create a series of 'village centres' in the area. This is a profound shift away from seeing north Solihull as a monolithic housing estate and instead trying to repackage and rebuild it into a series of distinctive settlements. Given the physical isolation of the site both from Birmingham and Solihull centres, this notion of villages has a great deal of appeal and draws somewhat upon the model of the polynucleated settlement. The villages will create walkable, mixed-use communities with greater accessibility of services and some forms of local employment. This is clearly more in tune with planning discourses of mixed-use, sustainable developments and does not draw on the central city model of small apartment development.

In terms of how this project is being carried out, the North Solihull Partnership has engaged in an extensive and sophisticated programme of involving the local community. A series of public consultations have been undertaken in each of the villages with the overall masterplans for each area subsequently revised in accordance with some of the recommendations by local residents. By drawing on local knowledge, some homes that might otherwise have ended up on the demolition list have been retained because they are actually popular with people living there. While this kind of process will inevitably have winners and losers, Solihull have not felt the pressure to deliver 'early wins' by demolishing and rebuilding areas without effective local consultation.

Key points

- Social housing suburbs are key targets for regeneration beyond the city core, with many such areas suffering from the indicators of social deprivation.
- While apartments are more culturally accepted as family housing in Scotland than in England, simply applying the metrocentric model of small flats will not produce demographically mixed communities.

> *(Continued)*
>
> - Locally available shops and services are crucial to attracting wealthier residents to regenerated areas and reducing car dependence.
> - The introduction of village-style community hubs is one mechanism for providing identity and coherence in the redevelopment of very large social housing estates, reproducing the suburban town model.

Ex-urban regeneration

The idea of moving urban development beyond the boundaries of existing towns and cities is not a new one. Britain's post-war new towns were predicated on the notion that existing towns and cities were already too big meaning that people and employers should be relocated from overcrowded urban areas to new settlements specially built some distance away. The new town model was seen as being exceedingly expensive and nowhere near the number planned in the 1940s were actually built during the 1950s and 1960s. There has been some discussion recently of reviving the model, Prime Minister Brown having pledged to build five new 'eco towns' each housing 10,000 to 20,000 people in low carbon emission homes (Watt and Revill, 2007).

The point of the new towns was that they should be independent, rather than a satellite of the old cities. A major criticism of recent development which has taken place at one remove from the city is that it retains functional dependence upon the city for jobs and services. The sustainability of such developments needs to be questioned as they tend to increase the reliance on the private car. These kinds of development are described here as being ex-urban – beyond the town. In some cases, these may be within the political control of city authorities, but are perceived as effectively being detached settlements. The first two case studies described here, indicate some of the problems that are generated by car-dependent, wealthy ex-urban developments, while the third suggests ways of overcoming these disadvantages.

Port Marine

Portishead is situated on the Bristol Channel about 13km west of Bristol and 30km from Bath. The town's two power stations fell into disuse during the 1980s, with neighbouring docks and factories similarly declining, producing a large waterside brownfield site for redevelopment. Being within 5km of the M5 motorway, Portishead is strategically well located within the economically dynamic south west region, making these brownfield sites a prime redevelopment opportunity. Outline planning permission was granted in 1997 and the former dock district has been branded Port Marine, with the lead taken by Crest Nicholson, a major UK developer which has focused its core business on regeneration activity (Figure 7.1).

In 1992, 13,000 people lived in Portishead, but when Port Marine is completed, it is expected that the population will be 30,000 with the development contributing some 4,000

Figure 7.1   Though a large-scale development, Port Marine has created a very attractive and varied new waterfront for Portishead. This brownfield development is anchored by a Waitrose supermarket, giving some indication of the wealthier social demographic being targeted.

new homes. This is a major increase to a small town, but where the post-war new and expanded towns were predicated on the idea that they would generate their own source of employment for residents, Port Marine has been primarily marketed at people wanting access to Bristol and the strategic motorway network in the region. As such, the development can be considered ex-urban as it looks outwards to other urban areas beyond the envelope of the town in which it is situated.

When developing Port Marine, Crest Nicholson avoided going down the route of reproducing the metrocentric, small apartment model and have provided a mix of three- to five-bedroom houses alongside apartment blocks. The development is not unproblematic, however. Indeed, the risk with a development of this kind is that where larger local authorities have more experience in squeezing concessions from developers – for example, on affordable housing, or contributions to community facilities – smaller authorities may find they have less bargaining power. Similarly, with a new residential population that tends to look beyond the town, the development may not contribute as much to the economic and social well-being of the host town as its size might suggest. There have been some rumblings of discontent even among the newcomers who moved to Port Marine that promised social facilities and public realm improvements have not yet materialised. The town also now has major traffic problems, particularly on the key A369, which links the town into Bristol, but there is no prospect as yet of reopening the old rail link that once served Portishead (Anon., 2007).

Nonetheless, Portishead is interesting because it indicates that developers see the possibilities offered by towns which, in and of themselves, would struggle to attract a large residential population. Even following its dramatic expansion, Portishead is still a small town, in pleasant rural surroundings with, as a result of the redevelopment, an attractive waterfront. Its strategic location is critical, however, because developers like Crest Nicholson would not look at, for example, an isolated village on the west coast of Scotland as a major development opportunity. With the Barker Report on land use planning (Barker, 2006) suggesting a strategic release of green belt and greenfield sites in some areas, developments of this kind may well increase following the lines of major transportation corridors. While Port Marine itself is far from being a bad example of regeneration activity, when considering its impact on the regional transport infrastructure, it is clear that if there is to be an increase in these kinds of ex-urban development, more thought will have to be given at government level to strategic transport improvements.

Waterfront Edinburgh

Waterfront Edinburgh reiterates the problems of not clearly considering transport issues before development takes place. The site is itself within the political boundaries of Edinburgh, but the nature of the development means that it can be considered as ex-urban rather than suburban. Historically, Edinburgh was not located on the waterfront and it was neighbouring Leith that was the main port of the Firth of Forth. Leith was absorbed into Edinburgh as the city grew during the inter-war period, but the area retains a detached feel, with the capital turning its back on the waterfront. The decline of the port has produced opportunities for a whole series of **brownfield** developments in Leith docks and further west along the waterfront. In 1999 a masterplan was drawn up by Llewelyn-Davies Architects for the area around Granton Harbour. This masterplan was subsequently adopted by Edinburgh City Council in 2001 as the development framework for 'Waterfront Edinburgh'. Although this area is only 4km from central Edinburgh, it has a distinct identity, while poor transport connections make it somewhat isolated.

The Waterfront Edinburgh masterplan had a number of key objectives:

- To deliver a comprehensive and viable regeneration plan to reinforce Edinburgh's role as a major international city.
- To produce a high-density live/work environment to produce a 'buzz' in the area.
- To socially and physically integrate the development with neighbouring communities and contribute to their regeneration (Edinburgh City Council, 2007).

Clearly, although the development as seen as part of the wider regeneration of Edinburgh, there is an emphasis on the area having a distinctive character and function, rather than simply being another one of Edinburgh's suburbs. North Edinburgh is rather a deprived area with a great deal of social housing. Unlike the plans for a mix of socially rented and private accommodation in Glasgow's Dalmarnock redevelopment, the development around Granton Harbour has been very much driven by the private sector. The development does sit in stark juxtaposition against the surrounding social housing and it is unclear how proximity to Waterfront Edinburgh will help regenerate neighbouring estates.

The emphasis on high-density living/working to create a 'buzz' is also interesting. There are clear parallels here to the metrocentric model of development, attracting busy young professionals without children to live in this area, but here without the major sources of professional employment that locate in the central city. While some of the new residents will doubtless work in the new business parks built as part of the development, it is clear that many of the people who move to the area will still be dependent on the city centre for employment.

The masterplan covers 57 hectares, which is a large site, but the somewhat deprived demographic of the surrounding areas suggests that it may be difficult to generate the kind of critical mass to give the waterfront district real independence from the central city. A new road has been built to service the developments and a new tram line is planned, linking the city, Leith, the new waterfront district and the airport. Formal approval for the city's tram network was granted in 2007 in spite of opposition from the Scottish National Party, which had formed a minority administration in the Scottish Parliament earlier that year and had a manifesto commitment to scrap the scheme. Even on the most optimistic assessment, the first phase of the scheme is not due to open until 2011. More worryingly, the link out to the waterfront is likely to be the first part to be dropped if the project needs to be scaled back for financial reasons (Ferguson, 2007). For the time being, therefore, the city's waterfront developments remain heavily dependent on the private car, putting additional pressure on to the already heavily congested road network heading into central Edinburgh.

There have been some local objections that the reconstruction of Granton has resulted in an entirely soulless area, a product of bland architecture and a complete lack of local services. In the 2005 nominations for *Prospect Magazine*'s annual Carbuncle award for bad architecture in Scotland, one local wrote of Granton:

> Have the people who have done this taken into account the street plan that connects the area to the city? Sticking in a tram-line merely highlights the fact that this area is as cut off from the city's street fabric as it ever was. The old village area on the shore front lacks a community feel like Leith. And the post-war housing estate behind hasn't had a single planning improvement made to it since its disastrous beginnings. When the current planned developments go ahead its going to look like the worst areas of London – deprived ghettos looking out on a wealthy one. (*Prospect Magazine*, 2006)

The 2005 award went to Cumbernauld, a dismal high water mark of 1960s modernist awfulness, but the fact that this brand new, very large redevelopment site should be among the final nominations is a sad reflection on the project. The phrase 'ghettos' is very telling of the project's isolation beyond a socially deprived suburban fringe. Although the views across the marina into the Firth of Forth can be very beautiful, the development appears to miss some of the key aims of contemporary regeneration – integration, mixed use and sustainability.

There have been some quite innovative things undertaken at Granton. Part of the site comprised the seventeenth-century Caroline Park, which had disappeared under subsequent developments, and there has been some attempt to recreate this. Similarly, some of the street layouts have sought to create interesting aesthetic effects with the underlying topography. Ultimately, a major part of this project's appeal is its waterfront status, but the fact that Edinburgh was not traditionally a port has meant that the waterfront is some way outside the centre, in contrast to the situation in neighbouring Glasgow where the Clyde runs through the city. Although on the fringe of the city, the waterfront developments in Edinburgh do appear to be completely dependent upon it, meaning that for all of the potential of this site, at the moment it is a large, car-based commuter settlement with some additional business park development. As such the development highlights the many of the potential pitfalls inherent to ex-urban projects.

Lightmoor

The final ex-urban case study gives some indication that such developments can be undertaken in a sensitive fashion that pays more than lip-service to ideas of sustainability and mixed development. Lightmoor is being built just beyond the urban fringe of Telford, which was itself originally a post-war new town built to take the overspill of people and businesses from the overcrowded West Midlands. Where Waterfront Edinburgh has been constructed as a middle-class outpost, the construction of Lightmoor is a partnership between private sector developers and Bournville Village Trust, a major housing association in the region. The project also has significant input from English Partnerships because the site was part of the land bank it inherited from the old Commission for New Towns. The deep pockets of English Partnerships and the social mission of Bournville Village Trust have produced quite a unique settlement. Building work began in summer 2005 on a 72-hectare greenfield site, with a projected cost of £31m. Certain historic features on the site, including hedgerows, lanes and parts of a canal, have been integrated into the design in an attempt to give the new development some character.

Planning permission was granted for 800 homes, of which at least 25% will be affordable, provided and managed by Bournville Village Trust. Permission was also granted for a primary school, community centre and a small amount of local retail suitable for a 'village'. The guidelines underpinning the development include:

- a well-defined compact village surrounded by landscape;
- a strong distinction between the recreational open spaces encircling the village and the protected rural wildlife site;
- a mixed-use centre arranged around the High Street and a village green, located so that foot access is promoted;
- higher residential densities clustered around the Village Centre, with areas of lowest density at the edges of the village where transformations between urban and rural character are made;
- the character of the existing lanes is retained and they are integrated into the movement network as recreational routes for pedestrians and cyclists (Lightmoor, 2007).

Essentially, what is being produced is a planned village with walkable local services and a mixed demographic, which ties in very closely with the new urbanist agenda. The development is something of a modern Bournville in that it is a planned settlement, beyond the urban fringe, with a strong sense of social mission. Technologically, the development is quite advanced and even before the government set targets for low carbon domestic buildings in 2006 the decision was taken that all houses in the development should meet EcoHomes 'excellent' standard. The buildings are also designed to be flexible, so that houses can be altered and extended as future needs arise, ensuring that the buildings have

Figure 7.2   Lightmoor, just outside Telford, has been designed with a similar density of housing and narrow winding streets as at Poundbury, though without the same degree of *faux*-historic buildings.

a longer life span. Other environmentally-friendly features, such as sustainable drainage systems (SuDS) for surface water runoff, have also been integrated into the design of the development.

Bournville Village Trust has a very good 'brand' in terms of social mission and is known for being able to deliver high-quality developments. This reputation is important when putting together a project of this kind, which would otherwise seem comparatively high risk to private developers because it is so innovative. The Trust also has a longstanding commitment to use innovative environmental technologies, having experimented with orientating houses to let in maximum sunlight as early as the 1920s. This reduces the costs of heating houses and has become a central principle used by contemporary sustainability gurus ZedFactory, who used solar orientation for their BedZED development. Lightmoor does, however, have the feel of a demonstration project, rather like Poundbury discussed in Chapter 6. Indeed, even the design of the new development has some parallels with Poundbury, with high-density building cover, attempts to subordinate the car and a post-modern pastiche of historic building styles (Figure 7.2). This said, Lightmoor does respond to its location beyond the urban fringe, producing a modern version of the rural village. Unlike Edinburgh Waterfront, Lightmoor cannot be accused of trying to apply the metrocentric model to a location far from the central city and does seem much more in tune with contemporary policy on questions of social integration and environmental sustainability. This said, the village is likely to remain for the most part a commuter settlement for people working in Telford and so significant questions remain about its reliance on the private car by virtue of its ex-urban location.

Key points

- Ex-urban developments exist on or beyond the fringe of urban areas and though detached from them retain a functional dependence upon the city.
- Where public transport infrastructure is not of a sufficiently high standard, ex-urban developments can put serious pressure on already heavily loaded road networks.
- Without giving careful thought to the provision of affordable housing, these kinds of development can end up being somewhat monocultural and contributing little to the regeneration of neighbouring areas.
- Attempts to produce a specific response to an ex-urban site, such as the more self-contained village model at Lightmoor, appear to be more closely allied to policy discourses on mixed use and sustainability.

Scaling up: mega-regeneration in the Thames Gateway

The final part of this chapter moves beyond the scale of small developments in the suburbs or just beyond the urban fringe to consider the reconfiguration of entire regions. The Dalmarnock development discussed above is considered to be a small part of the much larger Clyde Gateway development. This 20-year, £1.6bn project stretches across the boundaries of both Glasgow City Council and South Lanarkshire Council. The aim is to produce 10,000 new housing units, 21,000 jobs and a population increase in the area of 20,000. This is an exceedingly large development, which will have a dramatic impact on the river corridor running into Glasgow from the south and east (McLaughlin, 2007).

Glasgow Gateway is, however, quite a modest scheme in some regards in that it has a realistic timetable and does not attempt to completely alter the region to the east of Glasgow, but rather represents a coordinated series of planned additions to the existing built-up area. This is in stark contrast to what is being attempted on the river corridor to the east of London. In 1995 the Thames Gateway Planning Guidance Framework identified this region as a key location for new development. This is a truly massive area, consisting of some 100,000 hectares running east along the River Thames from Canary Wharf in London to Margate on the coast (Figure 7.3). The plans for this area were given weight by the Sustainable Communities Plan of 2003 where the government set out a strategy to build 200,000 new homes in south east England by 2016, 120,000 of which would be in the Thames Gateway. To put this development in context, just in terms of housing alone, the project is around twelve times bigger than that conceived for the Clyde Gateway. The Thames Gateway is, therefore, truly mega-regeneration.

This area has been targeted for this major development for three primary reasons:

- It contains 3,000 hectares, or one-fifth, of all brownfield land in south east England.
- It is located in the area of the UK with the highest demand for new housing.
- It is argued that the Thames Gateway has the potential to link London and the regions to Europe (ODPM, 2005a).

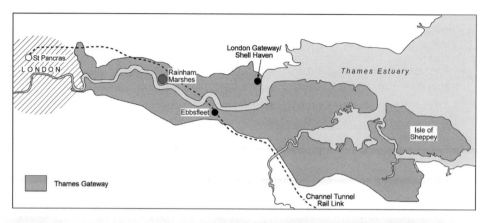

Figure 7.3 The Thames Gateway area, showing major developments. The Thames Gateway is a massive regeneration area that covers a range of land use types and jurisdictional areas. Its size raises serious questions about whether it is possible to regenerate an area of this scale in an integrated manner. Figure 7.3 derived from Department for Transport (2006), drawn by Kevin Burkhill.

In light of these factors, the government believes that the area has major economic potential and alongside the 120,000 new homes, it is hoped that it will create up to 200,000 jobs. As a result, the Thames Gateway represents the largest and most ambitious regeneration project in the UK (and, many claim, western Europe). But while characterised by similar policy goals to urban regeneration, it is possible to identify a number of distinct challenges, related to the scale of the proposed development:

- Strategic – the challenge of planning across a large area that contains urban, suburban and rural land uses.
- Governance – the organisational challenges of coordinating and delivering physical and social infrastructure on such a scale.
- Sustainability – the environmental challenges associated with building a massive new development in a floodplain.

The idea of undertaking regeneration on such a grand scale is clear – to allow greater coordination and integration of the different elements of development. However, making this work in practice is another matter. The National Audit Office (2007) issued a damning report on the progress of the Thames Gateway, claiming that very little had been achieved in the 12 years since the project began. These challenges will be examined in turn, considering the proposals themselves and the problems that have been encountered.

Strategic

The government's vision for the area is to create 'a world-class environment' (CLG, 2007b). There is no doubt that the scale of the project presents an opportunity to implement cutting-edge design and planning principles in order to create a new conurbation that could act as an example of best practice for the rest of Europe.

As Figure 7.3 shows, the Thames Gateway area includes a series of pre-existing urban areas, such as Barking and Ebbsfleet, and is already home to some 1.45 million people. Its boundary reflects the area alongside the Thames that was previously home to a number of industries, whose decline has left a legacy of dereliction and contaminated land. Covering 15 different local authorities, it contains some of the most deprived wards in the country and is characterised by a lack of services, unemployment and poor housing provision.

The strategic vision is to enhance existing urban conurbations and develop brownfield land, marshland and farmland across the sub-region, investing in major transport infrastructure to create a well-connected network of cities. The policy rhetoric framing the project mirrors that of urban regeneration more widely, aiming to attract business, provide high-quality housing, and improve the environment. The Thames Gateway's three areas – London, Kent and Essex – are complemented by four specific development sites, identified by government, which are referred to as 'transformal locations' (Thames Gateway, 2007). These four locations represent the focus of regeneration activities and are similar to the Area Based Initiatives (ABIs) discussed in Chapter 4 but on a larger scale. They are intended to drive growth in the surrounding areas and are located where there is a concentration of available land. It is worth considering three specific developments to understand how this process is intended to work.

One of the key economic developments is the London Gateway Development, a collaboration between DP World and Shell, to construct an international port and establish a major logistics and business park. The London Gateway Port will be located at the existing 600 hectare Shell Haven oil refinery in Thurrock and will be capable of handling the largest container ships in the world. Current plans anticipate that the port will be operational by 2011, with an attached business park opening at the end of 2008. The business park, known as London Gateway Park, will cover 300 hectares and aims to attract the usual high-tech sector, but also form a cluster for the distribution and logistics industries. The development itself is intended to create up to 16,500 new jobs and, more importantly, to act as a growth pole for the Thames Gateway regeneration initiative in Thurrock (Hammerton, 2005).

A number of major mixed-use developments are also planned. For example, the area around Ebbsfleet International Station (some 790,000 m²) adjacent to the Bluewater shopping centre is intended as the site for a development of housing, retail, residential, hotel and leisure sites. Major developer Land Securities (2007) plan to build, 10,000 new houses, five schools, leisure uses and transport links as well as setting aside 40% of the site as open space to create a sustainable community.

The East End of London also forms part of the Thames Gateway area, and the 2012 Olympic Games and Paralympic Games in London are being partially justified in terms of their contribution to these wider regeneration goals. Speaking at the Thames Gateway Forum in late 2006, Olympic Delivery Authority Chief Executive David Higgins described London 2012 as the 'Regeneration Games' (Olympic Delivery Authority, 2006). The area around the Olympic park will see the creation of 40,000 new homes, schools, health facilities and the largest new urban park in Europe for 150 years, with a network of restored waterways and wildlife habitats. As mentioned in Chapter 6, sports-led regeneration is often thought of as a catalyst for addressing existing economic and social problems, although the extent to which some of these developments would have happened regardless of the Games is open to question.

Governance

As discussed in Chapter 3, delivering integrated urban regeneration projects involves form-ing partnerships between a range of organisations from the public and private sectors. The task of coordinating these partnerships is amplified massively when regeneration is scaled up to a project the size of the proposed Thames Gateway development. Increasingly, the complexity of these partnerships is being blamed for a lack of coherent action on the Thames Gateway.

Taking economic regeneration as an example, the Thames Gateway development area falls under the jurisdictions of three regional development agencies (RDAs): the London Development Agency (LDA), the South East England Development Agency (SEEDA) and the East of England Development Agency (EEDA). Each of these RDAs has set up a sub-regional agency to attract inward investment to their respective parts of the Thames Gateway. 'Gateway to London' deals with the Thames Gateway London area, 'Locate in Kent' is responsible for the Thames Gateway North Kent area, and the 'Thames Gateway South Essex Partnership' handles queries relating to the Thames Gateway South Essex area. The government body UK Trade and Investment also has an 'Invest in Thames Gateway Team', which aims to 'progress the region's international agenda' for competi-tiveness, and English Partnerships are heavily involved. The development of the area is dependent upon the involvement of the private sector, but the complexity of the network of agencies, public bodies and partnerships involved has created confusion among would-be investors, acting as a barrier to regeneration (National Audit Office, 2007).

The confusion continues with physical and social infrastructure, as there are some 30 coordinating bodies involved, ranging from the Highways Agency to the Strategic Health Authority. The number of new agencies and partnerships continues to grow but progress often seems to be relatively slow, with individual developments taking place in isolation. The Thames Gateway Forum is the annual meeting place for all those involved in the regeneration of the Thames Gateway and Olympic region. In 2006 the event included 160 top-level speakers, and was billed as 'the largest ever gathering of the people and organi-sations responsible for delivering the most exciting regeneration project in the world' (Olympic Delivery Authority, 2006). Although the government has spent £7bn in the area since 2003, the National Audit Office (NAO) report found that ministers still do not have a single costed plan for the programme to join up local initiatives (National Audit Office, 2007).

An example of this is seen under health provision, where the Thames Gateway website only lists the new specialist cardiothoracic centre in Basildon, the Gravesham Healthy Living Centre and the Boleyn Medical Centre in Newham – no major new hospital is planned for the area. In terms of transport, the only major new infrastructure project is the Channel Tunnel Rail Link, which opened in late 2007, creating a high-speed railway line from London through Kent to the Channel Tunnel. It is intended to open a high-speed ser-vice that will link Ebbsfleet Station to London St Pancras in 15 minutes by the end of 2009, but beyond this the plans have been criticised, characterised as little more that the open-ing of a few new bus routes. Sir Terry Farrell, the architect responsible for, among other things, the MI6 building, said that the government seem to 'have handed out the jigsaw pieces, but there is no picture on the box' (Cavendish, 2007: 17). Unlike the Olympic Development Authority, which has a clear deadline by which it must deliver the venues and infrastructure for the Olympic and Paralympic Games, the wider Thames Gateway development has no deadline for completion.

A number of implications can be drawn from the institutional impasse that the Thames Gateway project appears to have run into: it may be that the area is simply too large to be planned in an integrated way, or it may be that the dominant models of urban regeneration are not suited to being scaled up on this magnitude. The partnership approach, for example, may become so complex that no action is possible when so many organisations are involved. Equally, the logic of private investment alone may not be capable of delivering the massive new infrastructure projects that are needed to create a new metropolitan region. Whichever explanation one chooses to accept, the problem of scaling regeneration up to the regional level remains pressing, and it is possible that radically different political bodies are needed to make regeneration work on this scale.

Sustainability

These problems are exacerbated by the threat of environmental change that will demand large-scale strategic responses. Decisions about where and how development takes place must be taken in the context of sustainability in order to minimise negative environmental impacts and maximise social and economic benefits. Major developments in the Thames Gateway will be seen as a blueprint for the government's approach to housing over the next 15–20 years.

Two key elements of sustainability are worth considering in the context of the Thames Gateway:

1   Issues associated with building in a floodplain.
2   Environmental protection.

Many of the new homes in the Thames Gateway area will be built in flood risk locations, adding to the number of properties currently at risk of flooding in the UK. The wisdom of this can be questioned, given recent high-profile debates concerning the problems of obtaining home insurance in areas prone to flooding, and the government's Planning Policy Statement 1 which places sustainability at the heart of the planning system. It has obviously been deemed that the area's advantageous location outweighs these considerations. Given the scale of the development, the problem of flooding associated with sea-level rise and increased storminess could be severe, and they are worth briefly considering.

Surprisingly, the Thames Estuary is one of the least understood and researched estuaries in the country. The current tidal defences for the estuary were built in the 1970s to protect against 1:2000 year flood (or a 0.05% risk of flooding). With sea-level rise, this will gradually decline, as planned, to a 1:1000 year flood (or 0.1% risk of flooding) by 2030. Over the past two years the region has experienced some of the worst floods on record. One of the worst affected areas was Kent, where 310mm of rain fell in October of 2002 compared to a monthly average of 80mm.

In light of these risks, reducing the environmental impact of homes in the UK is critical to making the Thames Gateway development sustainable. The local impacts of urban expansion are huge, and it is imperative that developers build to the standards set out in the new *Planning Policy Statement 25: Development and Flood Risk* (2006). For example, 2,250 new homes are planned within the tidal floodplain of the Swanscombe Peninsular (Kent Thameside Delivery Board, 2005). There is a need to protect the line around the western edge of the site, possibly through staging a managed retreat. In terms of heavy rainfall and runoff from impervious surfaces such as concrete and tarmac, much of the land has been raised by spoil deposition, which minimises the potential for flood storage as the soil is contaminated. Given the scale of the development, the existing system would not cope. The development thus requires the developer to contribute to the removal of material in order to provide flood storage.

The Thames Gateway area also includes rural and natural areas, such as the North Kent Marshes, which are recognised as Environmentally Sensitive Areas and Sites of Special Scientific Interest. Concerns have been raised because the Thames Gateway project threatens to develop significant areas of the marshland habitat. For example, the government's White Paper on air transport (Department of Transport, 2003) originally included proposals for a large international airport to be built on Cliffe Marshes in Medway. These were eventually dropped in 2003 due to opposition from local residents, the council, and various environmental non-governmental organisations. The expense of the plan was also deemed to be prohibitive, as it would have involved raising the ground level by 15m, but the government is looking at other potential locations for an airport in the area, including the feasibility of a floating airport near the Isle of Sheppey.

At the same time, a 600-hectare conservation park called Wildspace is being created at Rainham Marshes, in the heart of Havering. At three times the size of Hyde Park the unprotected marsh and landfill area will become a wildlife sanctuary. But controversy reigns over the decision to spread the Thames Gateway development over such a large area. Many commentators have suggested that new housing should be confined to the area within the M25, and that more high-quality greenspace is needed in order to lure an upwardly mobile professional workforce into the area. In terms of the marshes, there is a danger that planners will simply 'ring-fence the best and trade-off the rest' for development (Selman, 2002: 284).

Such tensions are inevitable when considering a development on the scale of the Thames Gateway. Environmental impacts are unavoidable, but it is important that the social goals of the project are not frustrated at the expense of poorly conceived plans. For example, the government is currently considering whether to build the £450m Thames Gateway Bridge in east London. The public inquiry showed that the proposed six-lane urban motorway would cause major local air pollution and congestion that would impact most heavily on exactly those poor communities that the Thames Gateway project aims to help. As discussed in Chapter 5, questions of who will benefit and who will lose out characterise many regeneration projects and it has been suggested that the dual goals of sustainability and community cannot be delivered within the neoliberal framework of the Thames Gateway development (Edwards, 2008). The scale of the Thames Gateway makes these issues harder to ignore and harder politically to resolve, as local opposition to parts of the strategy can delay or damage the overall strategic goals of the development. It is within contexts such as this that the recent White Paper on Planning, which sought to remove the ability of local groups to block large-scale developments, can be understood.

Key points

- The principles of regeneration are being applied on increasingly large scales.
- The Thames Gateway has clear parallels with urban regeneration being characterised by numerous brownfield sites and deprived communities.
- Proximity to London makes the Thames Gateway a key strategic site to ease problems of housing supply in the south east in spite of the problems associated with development on a floodplain.
- Recent reports suggest that is hard to coordinate effective regeneration partnerships across such massive areas.
- Ensuring that developments on this scale adhere to the principles of sustainability involves making major political trade-offs.

Conclusion

The Thames Gateway development will eventually function as a polynucleated settlement, with a series of urban hubs within a region which retains a close dependence on neighbouring London. The scale of the development is perhaps proving too much for existing mechanisms of regeneration governance in the UK. Nonetheless, with the relaxation of restrictions on greenfield development proposed by Kate Barker, it seems likely that there will be an increase in new construction at and beyond the urban fringe.

As with development at the regional scale, ex-urban development needs to be carefully managed. Simply giving private developers the opportunity to build large numbers of houses will not deliver the kinds of local facilities that will be needed to reduce car dependence. Similarly, if development is going to take place in this fashion, these areas will function as satellites of established urban areas – something which the post-war concept of new towns was explicitly attempting to avoid. There are profound implications for transport and increased congestion if this kind of development is not carefully managed.

One of the major targets for regeneration in existing suburbs is the redevelopment of large social housing estates. Attempting to reproduce the suburban town model has some virtue in these areas as the provision of local services and a local identity may help to attract and integrate a wealthier population to break down the concentration of socially deprived individuals in these areas. The idea of suburban towns – essentially the polynucleated model applied within the existing city – provides a good compromise in terms of sustainable development. These areas function best where there are good public transport links to the central city, for specialised services and employment functions, but also some provision of retail and other functions locally at a walkable scale. These kinds of development are perhaps even more vulnerable to competition from the car-dependent, out-of-town retailers than city centres, indicating the importance of thinking very carefully about the wider impact of developments on the urban fringe.

What is clear, however, is that in the UK the suburbs are still seen as a key location for families with children. While the model of small apartment development has worked spectacularly well in city centres, it has catered to a very specific, young demographic and attempts to apply this metrocentric model to suburban and ex-urban developments holds some real risks. The suburban experience is one which is qualitatively different from that in the central city and it is clear that to apply city-centre models in outer urban areas will not be sustainable either in terms of mixed communities or in attempting to reduce car dependence. The relatively small amount of research and policy that responds specifically to the challenge of regeneration beyond the city core indicates that these issues have not yet been carefully thought out. The severe difficulties encountered in the Thames Gateway regeneration are perhaps symptomatic of this broader problem.

Further reading

Suburban redevelopment is a fascinating topic which has not really had the specific atten-tion that it deserves. Nathan and Unsworth's (2006) article is good introduction to the tension implicit in attempting to apply city-centre models of regeneration to suburban areas, although their work focuses specifically on the inner suburbs. The report by the In Suburbia Partnership (2005) gives some indication of how the challenges of suburban regeneration might be tackled. The special issue of *Urban Studies* (2001) provides a rigor-ous exploration of the complex ideas underlying notions of the polynucleated city and region. The National Audit Office (2007) report provides a good insight to the challenges facing those attempting to deliver the regional-scale transformation in the Thames Gateway as does the edited collection by Cohen and Rustin.

In Suburbia Partnership (2005) *In Suburbia: Delivering Sustainable Communities* (Civic Trust, London).
Cohen, P. and Rustin, M. (eds) (2008, in press) *London's Turning: the prospect of Thames Gateway*. London: Ashgate.
Nathan, M. and Unsworth, R. (2006) 'Beyond city living: remaking the inner suburbs', *Built Environment*, 32(3): 235–249.
National Audit Office (2007) *The Thames Gateway: Laying the Foundations* (HMSO, London).
Urban Studies (2001) 'Special Issue: The Polycentric Urban Region' *Urban Studies*, 38(4).

Note

1 In the context of post-war council housing, the maisonette was a distinctive type where two-storey flats were stacked one on top of another. These flats were specifically designed for families. Four- and six-storey blocks of this type were built without lifts, making them a nightmare for the elderly or those with small children.

8 | Conclusions

Challenges revisited

At the start of this book we discussed the key issues that urban regeneration seeks to address. It is worth briefly reconsidering these in order to contextualise the themes that emerged in the subsequent chapters. Initially, regeneration was seen as a way to reverse the decline of industrial cities associated with the loss of manufacturing industry over the second part of the twentieth century. Industrial decline had led to severe dereliction and depopulation in certain areas, creating a series of social and economic problems, particularly within inner cities. A key question exercising urban regeneration was how to make cities attractive places in which to live and work. More recently, the issue of providing adequate housing supply for the population has returned to the fore, as soaring house prices and increased levels of household formation have exacerbated shortages around the country. Rather than 'concreting over the countryside' with sprawling new developments, existing cities are seen as the ideal places in which to solve housing shortages by bringing derelict land back into use.

Given the massive scale of these challenges, it is easy to see why the urban regeneration agenda has assumed such importance in the contemporary political landscape. From Secretary of State for the Environment Michael Heseltine visiting Liverpool in the wake of the Toxteth riots in 1981, to the ongoing Olympic Games-orientated regeneration of East London, the driving force behind regeneration has been primarily political in nature. There are, however, a multitude of steps between issuing political imperatives and actually getting to the stage of laying one brick on top of another. A wide variety of people and agencies need to come together to make regeneration happen in practice. Regeneration requires input from diverse sectors, such as planning, development, health, environment, transport and education, with coordination and resources needed from both public and private organisations. Regeneration involves developers, government, communities, architects and planners working closely together. Partnerships and collaboration are required across different sectors in order to avoid the mistakes of the past and create quality, sustainable environments where people will want to live and work.

Key themes

The creation of sustainable environments is very tricky to achieve in practice. The preceding chapters have worked through various dimensions of this challenge, considering the ways in which regeneration has been undertaken and how policy and practice has evolved since the early 1980s. A number of key themes have emerged from this, and it is worth considering these as they constitute the defining elements of urban regeneration in the UK today.

Partnership

Perhaps the most dramatic difference between urban regeneration and previous interventions in UK cities has been the emphasis put on partnership between different branches of government, the private sector and communities. Partnership is now central to urban policy – it is no longer the case that a local authority can simply decide to rebuild a rundown part of the city and then make it happen. Because the political culture of the UK has shifted to a more **neoliberal** position, state funding for projects is seen primarily as a means to draw private sector investment to areas which otherwise would be seen as too risky or difficult. This said, a key critique of urban regeneration in the 1980s was that this practice of drawing in private sector partners simply acted as a state subsidy to wealthy developers. As the scope of regeneration has developed to bring together a social, economic and environmental component this critique has ebbed somewhat, with a whole variety of outputs being sought from the private and public sector investments.

Given the broadened remit of regeneration, there is a logic to bringing in a variety of actors with different expertise from the public, private and charitable sectors. Scholars have developed the idea of governance to help explore this bringing together of different actors. Partnership as a mode of governance is interesting as although non-state actors have been brought into the process, critics such as Jonathan Davies (2002) have argued that the aims of the state are still dominant. The state comprises a complex series of overlapping agencies operating at a variety of geographic scales which pursue different and sometimes contradictory aims. As a result, it is perhaps rather simplistic to reduce regeneration to delivering the aims of the state, particularly as the involvement of multiple partners can change the intended direction of a project. Furthermore, as the example of the Thames Gateway shows, the larger the project the more difficult it seems to be to coordinate the many different partners involved. Ironically, then, on the kinds of project which require most input from other agents because of their complexity and the number of people being affected, the current governance structures in the UK seem least able to cope.

Sustainability

Since the idea of sustainable development entered the mainstream in the late 1980s and early 1990s, it has now reached the point where it is ubiquitous in all policy initiatives. Sustainability and regeneration go hand in hand, both concerned with the interplay of social, economic and environmental needs. Ideas of sustainability underpin many of the key trends in regeneration activity in the UK, for example re-use of brownfield sites to reduce urban sprawl, bringing economic activity to declining regions and tackling social issues such as access to affordable housing.

Sustainability is, however, frequently about trade-offs. Economic activity, for example, can bring social benefits, but can also have negative impacts on the environment. Those undertaking regeneration schemes have to balance competing agendas, perhaps seeing ecologically diverse **brownfield** sites more for their potential to provide affordable homes than their habitat and amenity value. Getting this balance right can be difficult and, particularly in England, it is arguable that decision-making structures have swung too far towards economic agendas. In Scotland, Wales and Northern Ireland there tends to be more connection between regeneration and social policy in the responsible government departments. In England, social issues have tended to be shunted into the separate **discourse** of community 'renewal', while large parts of the regeneration agenda have been passed to the regional development agencies, which have a primarily economic remit.

There is also a danger with sustainability in that, because of its breadth, the concept becomes somewhat meaningless. No one working in regeneration in the UK today would willingly claim that their project was not sustainable. If, for example, a developer makes money selling apartments, then the project is, if nothing else, economically sustainable regardless of its effects on local communities and the environment. In spite of some cynicism about how the word is applied, one should not lose sight of the fact that sustainability is a very enlightened concept, pushing regeneration towards more **holistic** approaches and trying to make the world a better place.

Tackling industrial decline

Perhaps the single biggest factor driving regeneration in the UK has been the shift to a post-industrial economy. Large areas of towns and cities fell into dereliction as industrial functions moved elsewhere as part of the restructuring of the global economy. This also left a legacy of high unemployment and economic stagnation in many urban areas of the UK. Regeneration has sought to reverse the flow of higher income groups out of the city and attract new businesses and forms of economic activity. Services, tourism and leisure have become ever more important, meaning that cities can no longer function as places where people simply work before retreating to suburban enclaves each evening.

The transformation of former industrial sites and under-utilised portions of city centres has driven the regeneration process. Brownfield redevelopment has allowed urban economies to expand without further sprawl into the countryside. Areas of former dockland have become high-value office spaces, Victorian factory buildings have been converted into loft apartments. The economic disaster of industrial decline has, through regeneration, provided the opportunity for wholesale change and is a story of economic growth in particular sectors and particular parts of the country. One should not lose sight of the fact, however, that not everyone has benefited equally from new economic growth.

Knowledge and the new economy

Over the past quarter century the UK has seen a process of re-imaging being undertaken to make cities places where people actually *want* to be. Manchester, for example, has been transformed from being seen as a grimy, northern industrial city, to being hip, fashionable and dynamic – a place where people are excited to live. Regeneration has driven this transformation, producing new public spaces, new facilities, new apartments and new jobs in new sectors of the economy.

The knowledge economy has been at the heart of this change, with brains rather than brawn being the engine of growth. Not all regions and demographic groups have done well out of this economic restructuring, with low-paid service jobs replacing high-skill manufacturing work in many areas. There has, however, been a great deal of enthusiasm among policy-makers to attract to their cities Richard Florida's (2002) 'Creative class', working in the IT, media and communications sectors, through attempts to make a more attractive urban environment. There is also a belief that the development of these creative industries can be fostered through the establishment of clusters, where people working in these sectors can easily meet and network with each other, to add value to their businesses. A number of towns and cities have now attempted to produce creative quarters of one kind or another, some including cultural facilities, in order to anchor nascent clusters, for example the FACT centre in Liverpool's Ropewalks district. This does, however, raise a broader question about distinctiveness; if all towns and cities are pursuing similar strategies to try to attract creative businesses, what makes an individual town stand out in a competitive market?

City living

One of the most profound shifts in the functioning of UK towns and cities in the last quarter century has been the return of population to central urban areas. In the 1980s city centres were not places where people lived, while today exclusive flats and apartments in urban cores are fashionable, prime real estate, with many more being built each year. This has been accompanied by the rise of complimentary leisure functions, such that UK city centres no longer shut down at 5pm. This model of inner urban redevelopment has been immensely profitable for developers and ticks a great many boxes in terms of reducing car dependence and providing accommodation for the growing number of smaller households. The city centre is not the whole city, however, and it is possible to identify cases where an overly **metrocentric** model has been inappropriately applied to outer urban areas.

There has been a renewed emphasis on high-quality design as part of the attempt to attract people back into cities. From flagship architectural statements in the city centre, to the sensitive restorations of historic buildings elsewhere, to simple improvements in the form and legibility of the **public realm**, cities are being made more attractive places to live. This not only applies to city centres, but also to thoughtfully designed settlements elsewhere, making use of smart growth principles of walkability and high-density design alongside innovations such as design coding. It should be noted, however, that while there has been a great deal of innovative and high-quality design, particularly in the last ten years, there has also been an explosion of bland, characterless buildings, constructed to price by the major housebuilding firms. This kind of design does little to improve the city living experience.

Gentrification

The final key theme is perhaps the most controversial. The word '**gentrification**' is never used by those working in the regeneration sector, but almost all regeneration activity is predicated on attracting new people and new businesses to run-down areas. This undoubtedly makes it easier to bring private developers on board, who can market to a more wealthy demographic and, indeed, can allow public bodies to negotiate with developers in order to produce a proportion of affordable homes, subsidised services and other social benefits. Bringing in new people and businesses very frequently involves the displacement of existing residents – often some of the poorer and more vulnerable members of society.

The connection between regeneration and gentrification is an intractable issue and this book does not suggest that there are any easy answers. One of the reasons why this book has not discussed issues of community in any detail is because existing communities, in practice, tend to be by far the weakest actors in programmes to physically transform urban areas, for all of the policy rhetoric about inclusion and social cohesion. We do not live in an ideal world and the question of gentrification brings us back to the need for trade-offs. Those existing residents who are able to remain in a dramatically regenerated area should benefit from the increased value of their homes, an improved environment, better local services and a generally healthier local economy. Other residents will be priced or forced out of that area. There have been cases where artist communities have been deliberately used to make a low-rent area fashionable before being displaced by wealthier groups attracted to the very thing they are forcing out. One danger in the overlap of regeneration and gentrification is of producing very bland, monocultural developments, comprising young professionals without children housed in soulless, generic buildings. This may be the opposite of the policy rhetoric, but in the UK over the last decade, this kind of development has happened all too frequently.

Reflections on the future of urban regeneration

Like all forms of politics, urban regeneration is the 'art of the possible'. The overriding message that comes across from each of the chapters is one of compromise and trade-offs between different concerns. Successful urban regeneration schemes occur when collaboration has been effective and fair, while difficulties are related to failures to achieve equitable balances. One of the systemic challenges to regeneration involves ameliorating the excesses of the neoliberal approach, which has, in cases, resulted in a rather uneven distribution of economic benefits, and the sidelining of social and environmental concerns. Both Glasgow in Chapter 4, and Salford Quays in Chapter 5 question the extent to which those people in the greatest socio-economic poverty have benefited from urban regeneration. The prioritisation of economic development is detectable in the Sustainable Communities Plan, which seems to have few solid targets for social and environmental sustainability. It has probably been the most controversial element of the ODPM/CLG's work, focusing instead on the massive expansion of housebuilding in the south east combined with demolition (and gentrification) in the Pathfinder areas. More recently, the Treasury-driven Barker reviews have placed more emphasis on market forces determining how land should be developed and this has subsequently fed through into planning policy.

These observations are not intended to indict regeneration in the UK as some kind of failure. It is important to remember that the initial goals of regeneration were economic, and by most economic indicators prosperity in the urban areas of the UK has increased substantially. Cities still face challenges, however. As the recent *State of the English Cities*

Report (ODPM, 2006) claims, levels of socio-economic deprivation remain higher and more widespread in cities, reflected not least in higher levels of unemployment. Furthermore, in the face of future challenges, these tensions will become increasingly hard to ignore. By way of tying up the arguments of this chapter, it is worth finally reflecting on the future challenges and potential trends within the sector.

Not all cities can compete within the global economy to attract the most desirable industries, and individuality will become increasingly critical to the success of cities. It was perhaps excusable that planners in the 1980s and 1990s accepted relatively generic architecture and a preponderance of flats and apartments in order to kick-start regeneration and meet housing targets. Regeneration projects are often (erroneously) described as creating truly unique places and one way this can be achieved is through retaining a sense of history and local culture. Within the broader evolution of regeneration in the UK today, however, cities need to be more creative in terms of how they develop in order to differentiate themselves from other places. Cities can afford to be more demanding of developers, as many have a proven track record of profitable developments. Local authorities no longer need to beg developers to come and build anything they want simply to attract some form of investment. For example, rather than demolishing vast areas in order to present developers with 'attractive' (i.e. large) land packages, cities can opt to retain characteristic features, demanding more creative and higher-quality development proposals. This would also address issues of social inclusion and cultural retention, as existing communities would be included in the vision for an area.

The spectre of global environmental change presents possibly the greatest challenge to urban regeneration in the twenty-first century, and there are serious issues in terms of how planners and developers redress the balance between economic gain, the social imperative to build houses, and the necessity of environmental awareness. It will become increasingly difficult to site developments if flooding becomes more widespread, whether due to increased intensity or quantity of rainfall, or sea-level rise. Similarly, new developments will have to be built more sustainably and be more sustainable to live in. In terms of building and planning sustainable developments, much technology and knowledge already exists. The critical challenge is making sure that it is adopted. Most professionals are aware, for example, that building in floodplains is environmentally inadvisable, but this knowledge is ignored because floodplains represent a cheap supply of under-utilised land. Eco-friendly housing already exists in many forms, but the challenge of adopting it more widely is often misrepresented. Eco-friendly housing is more expensive and thus, it is argued, unsuitable for mass-production. Developers argue that the primary force driving the kinds of houses they build is consumer demand. Both UK house buyers and the volume housebuilding firms tend, however, to be rather conservative – it is not that the market does not exist, but rather that education is needed to help realise it. In this sense, urban regeneration is caught up in the wider politics of environmental change, as the government struggles to raise public awareness and create institutions capable of engendering joined-up thinking.

A broader political tussle in which urban regeneration is caught concerns the question of whether devolution and empowerment are actually occurring, or whether power is being centralised. On the one hand, regeneration policy has been profoundly affected by the introduction of a regional tier of government, the emphasis on local action embodied in sustainability and the neoliberal discourse of self-help and community ownership. On the other hand, the government has been widely criticised for not living up to this rhetoric of local empowerment and, in some cases, of even reinforcing the powers of Whitehall. The recent broadening of the powers on compulsory purchase and reducing the opportunities for communities to stop large projects deemed to be of national importance do little to counter this view.

This tension is not unrelated to the challenges of social and environmental change. Central government is undoubtedly entering an era in which it will need an executive capacity to respond to large-scale challenges, such as housing shortages and sea-level change. The UK has a medium-term problem with energy supply, for example, as a series of nuclear power stations reach the end of their life over the next 20 years. There are very large issues to be addressed about the siting of new power stations and encouraging measures to reduce household energy use. The location of such facilities can be a source of immense local controversy, as nobody wants these close to their homes. Large-scale developments like the Thames Gateway require massive **infrastructure** building programmes, to provide sustainable transport and waste systems. The recent planning reforms, which remove powers from communities to contest the siting of major developments that are deemed to be in the national interest, are intended to make it easier for the government to respond to major challenges with large-scale strategic developments. The need to balance increased executive decision-making powers with the empowerment of local communities represents a major political challenge that goes well beyond the regeneration agenda. Debates over the best way to govern in the twenty-first century, and issues surrounding social and environmental citizenship are all pertinent to the field of regeneration.

Finally, and in the face of the challenges outlined above, the importance of learning from international examples is paramount. Every week there are reports in the press of various innovative developments from around the world, whether it is Dongtan, the new eco-city being built in China to house 500,000 people, or the totally car-free Vauban development in a suburb of Freiburg, Germany, where 2,000 new homes have been built. Indeed, Prime Minister Gordon Brown unveiled plans to build eco-cities in the UK as part of a commitment to increase the supply of homes, although there is some debate over when and where they will be built. While regeneration in the UK has generally followed trends from the USA in the past, the sector is undoubtedly global now and many of the challenges facing the UK are the same as those faced elsewhere. In order to meet them in the most effective way it is necessary not only to innovate but also to learn from others. As this book has shown, the regeneration agenda has already had a massive impact on cities in the UK. Cutting across a wide range of activities, regeneration is one of the most dynamic and innovative sectors in UK policy and practice. The next generation of professionals choosing regeneration as a career are entering a fast moving world, with great challenges and tremendous possibilities.

Keeping up to date

Urban regeneration is a rapidly evolving field and it can be quite difficult to keep up with all the changes, which is why certain key sources are worth consulting on a regular basis. The names of government departments and their web addresses change over time and departments dealing with urban regeneration seem to be re-branded on a regular basis. Nonetheless government sites are very useful in giving access to key policy documents. It is often helpful to supplement these rather technical documents with the evaluations of them produced by newspapers and pressure groups, although one should always remember the political position from which journalists and lobbyists are working.

http://www.lexisnexis.com/uk/nexis   Those readers who are subscribed to this part of the Lexis-Nexis database have access to a text-searchable archive of UK newspaper articles which stretches back into the early 1990s.

http://www.regeneration-uk.com   Provides an excellent portal with links to a variety of sources for news and other resources, though perhaps with a more economic emphasis.

http://www.dsdni.gov.uk/   Department for Social Development in Northern Ireland

http://wales.gov.uk/   The main portal for the Welsh Assembly and Welsh Assembly government – the individual departments do not have separate sites.

http://scotland.gov.uk/   The main portal for the Scottish Government – the individual departments do not have separate sites.

http://www.communities.gov.uk/   Department for Communities and Local Government in England, with good links to relevant parts of other departments and executive agencies.

http://www.cabe.org.uk/   Commission for Architecture and the Built Environment has lots of good material on urban design.

http://www.rtpi.org.uk/   The Royal Town Planning Institute is the professional body for planners in the UK and has excellent coverage of the latest news and issues in the planning arena. It also has separate content specifically relating to Scotland.

http://www.riba.org/   The Royal Institute of British Architects is the professional body for UK architects and has good content on current debates in the world of design.

http://www.bura.org.uk/   The British Urban Regeneration Association, a not-for-profit body championing the interests of those involved in regeneration.

http://www.intbau.org/   The International Network for Traditional Building, Architecture & Urbanism is a global organisation which promotes local character and traditional architecture. Its patron is HRH the Prince of Wales and it can be seen as broadly new urbanist in outlook.

Glossary

Blairite: the adjectival form of Blair, who led the Labour government between 1997 and 2007. The word 'Blairite' refers to the supporters and policies of Blair's government. The hallmarks of Blairite policy include the increased use of markets to deliver public services, but reined in through partnership with the public sector, and pro-European and devolutionary policies.

Bond: a bond is a financial instrument that represents a debt security, whereby the bond holder lends money at a rate of interest that is repaid at the end of the term, or when the bond 'matures'.

Brownfield: refers to previously used land. The word 'brownfield' was coined in opposition to the term 'greenfield', which designates a development site in a previously undeveloped areas. It includes the categories derelict land, which is previously used, and contaminated land, which is previously used and polluted in some way. Brownfield is synonymous with the US term 'brownland'.

Capital: in a financial sense, capital is any asset that can be used or invested. It is usually taken to mean privately owned wealth. In recent times capital has also been used in the sense of human capital or social capital, to indicate the strength of social networks, in terms of shared interests and civic engagement.

Discourse: a discourse is a set of specific meanings or representations that are attached to certain things. So, for example, one discourse of inner cities represents them as dangerous and crime-ridden. Because different groups often represent things in different ways, there may be different discourses about the same thing. A contesting discourse of inner cities that is becoming more dominant is that they are vibrant and diverse places to live.

Equity: the principle of fairness between groups of people, often designated as a key principle of sustainable development. Equity can also mean the share of a person's ownership in an asset when used in a financial context.

EU: The European Union (EU) is a political body of 27 member states. The EU evolved from the European Community (EC) at the Maastricht Treaty in 1993. The European Commission represents one of its political bodies, and forms policy on regional development, agriculture and the environment, among other things.

Gentrification: the process by which buildings or residential areas are improved over time, which leads to increasing house prices and an influx of wealthier residents who force out the poorer population of an area.

Glocalisation: a term derived by combining localisation and globalisation, which highlights the idea of behaviour which is simultaneously acting to an increasing degree at both a specific local level and at the global scale ("act locally, think globally"). In the specific context of governance it can be used to refer to the hollowing out of the nation state, with powers increasingly passing upward to supranational organisations and downward to local communities.

Holistic: literally means addressing the whole. It is usually used to mean an integrated approach that considers all aspects of a problem.

Infrastructure: in the context of urban regeneration, infrastructure designates the 'hard' engineered features of the urban environment, including roads, water pipes, electricity, waste systems, railways, pavements, lighting, and so forth.

Managerialism: see New Localism.

Metrocentric: a term used to describe a focus on central city issues.

Mixed development: a general term to designate developments that include more than one kind of use. Usually, mixed developments include retail uses (shops), residential (homes), business premises (offices) and leisure uses (cafés, bars and so forth). The term is also sometimes used to designate a mixture of residents and users; this can include mixed tenure (i.e. some rental, some owner-occupied), mixed income groups and mixed ethnicities.

Neoliberal: an approach that believes markets provide the best solution to social problems. So, for example, the introduction of carbon trading as a way to curtail greenhouse gas emissions is a neoliberal policy response. The approach builds upon classic economic ideas developed by Adam Smith, and was championed by the New Right in the USA during the 1980s.

New Labour: term applied to Tony Blair's Labour government that took power in the 1997 general election. They were considered 'New' as they moved away from traditional left-wing policies (such as their long-standing affiliation with the Trade Unions) towards the centre ground, or so-called 'Third Way'. The idea that they were 'New' also articulated the fact that this was the first Labour government for almost 20 years, and that they were led by a young dynamic leader.

New Localism: is used to describe the tendency of Blairite policies to devolve the implementation of policy goals down to the local level. A key feature of this trend is the devolution of management to the local level in order to achieve policy goals more efficiently, although the political power to decide what those goals should be is generally not devolved. The New Localism is thus closely linked to the emergence of managerialism at the local level.

Procurement: the acquisition of goods or services for an organisation or individual at the best possible price. The EU public procurement directive requires public bodies to put all their procurements out to competitive tender in order to reduce corruption and ensure that the cheapest services and goods are obtained.

Public Realm: commonly used term meaning public spaces and activities. Sometimes applied to areas of policy that directly concern the public.

Quango: an acronym for QUasi-Autonomous Non- (sometimes 'National') Government Organisation. Quangos have proliferated under New Labour, as various powers and responsibilities of the state are devolved to organisations that are neither public nor private. Quangos have been criticised because they are not democratically elected and are often not accountable to the public for their actions. The field of urban regeneration is populated by many quangos.

Remediation: the process of cleaning up polluted brownfield land.

Thatcherite: the adjectival form of Thatcher, a Conservative party politician who was Prime Minister between 1979 and 1990. She was the first female Prime Minister in the UK and the longest-serving PM of the twentieth century. Her government was associated with introducing right-wing neoliberal policies from the USA and reducing the role of the state in providing for basic social needs like health and housing.

References

Adair, A., Berry, J. and McGreal, S. (2003) 'Financing property's contribution to regeneration', *Urban Studies*, 40: 1065–1080.

Aldrick, P. and Wallop, H. (2007) 'For sale: too many flats, not enough houses', *The Daily Telegraph*, 29 May: 1.

Alker, S., Joy, V., Roberts, P. and Smith, N. (2000) 'The definition of brownfield', *Journal of Environmental Planning and Management*, 43: 49–69.

Anon. (2007) 'Boomtown stats', *Bristol Evening Post*, 17 March: 6.

Armstrong, P. (2001) 'Science, enterprise and profit: ideology in the knowledge driven economy', *Economy and Society*, 30: 524–552.

Arnstein, S. (1969) 'A ladder of citizen participation', *Journal of American Institute of Planners*, 35: 216–224.

Atkinson, R. (1999) 'Discourses of partnership and empowerment in contemporary British urban regeneration', *Urban Studies*, 36: 59–72.

Atkinson, R. and Helms, G. (2007) *Securing an Urban Renaissance: Crime, Community, and British Urban Policy* (Policy Press, Bristol).

Babtie (2000) *Statement of Principles* (University Hospital Birmingham NHS Trust, Birmingham).

Bailey, N. (2003) 'Local Strategic Partnerships in England: the continuing search for collaborative advantage, leadership and strategy in urban governance', *Planning Theory & Practice*, 4: 443–457.

Bailey, N. and Turok, I. (2001) 'Central Scotland as a polycentric urban region: useful planning concept or chimera?', *Urban Studies*, 38: 697–715.

Barker, K. (2004) *Delivering Stability: Securing Our Future Housing Needs*. Final report – recommendations (HM Treasury, London).

Barker, K. (2006) *Barker Review of Land Use Planning*. Final report – recommendations (HM Treasury, London).

Bartlett, E. and Howard, N. (2000) 'Informing the decision makers on the cost and value of green building', *Building Research & Information*, 28: 315–324.

BBC (2006) Online map 'misses' regeneration.

Begg, I. (2002) *Urban Competitiveness: Policies for Dynamic Cities* (Policy Press, Bristol).

Bell, D. and Jayne, M. (2003) 'Design-led urban regeneration: a critical perspective', *Local Economy*, 18: 121–134.

Binnie, J. and Skeggs, B. (2004) 'Cosmopolitan knowledge and the production and consumption of sexualized space: Manchester's gay village', *Sociological Review*, 52: 39–61.

Birmingham City Council (2002) *Selly Oak Hospital* (Department of Planning and Architecture, Birmingham).

Bond, D. (2007) 'Costs "ruining" Games legacy', *The Daily Telegraph*, 16 March: 1.

Brennan, A., Rhodes, J. and Tyler, P. (1999) 'The distribution of SRB Challenge Fund expenditure in relation to local-area need in England', *Urban Studies*, 36: 2069–2084.

Bryson, J. and Buttle, M. (2005) 'Enabling inclusion through alternative discursive formations: the regional development of community development loan funds in the United Kingdom', *The Service Industries Journal*, 25: 273–288.

CABE (2000) *By Design: Urban Design in the Planning System: Towards Better Practice* (DETR, London).

CABE (2004) *Creating Successful Masterplans: A Guide for Clients* (CABE, London).

CABE (2005a) *Creating Successful Neighbourhoods: Lessons and Actions for Housing Market Renewal* (CABE, London).

CABE (2005b) *Making Design Policy Work: How to Deliver Good Design through Your Local Development Framework* (CABE, London).

Cameron, S. (2003) 'Gentrification, housing redifferentiation and urban regeneration: 'Going for growth' in Newcastle upon Tyne', *Urban Studies*, 40: 2367–2382.

Cameron, S. (2006) 'From low demand to rising aspirations: housing market renewal within regional and neighbourhood regeneration policy', *Housing Studies*, 21: 3–16.

Cardiff Harbour Authority (2007) Cardiff Bay Development Corporation (http://www. cardiffharbour.co.uk/learning/about_cbdc.htm, accessed 14 July 2007).

Cavendish, C. (2007) 'We're stuck at the Gateway to nowhere', *The Times*, 31 May: 17.

CLG (2006a) *Government Confirms Cash to Benefit Communities*. News release 20 June (Department for Communities and Local Government, London).

CLG (2006b) North East Objective 2 (http://www.erdf.odpm.gov.uk/ERDFFundedAreas/ NorthEast/NorthEastObjective2, accessed 20 September 2006).

CLG (2006c) *Planning Policy Statement 3: Housing* (CLG, London).

CLG (2006e) SRB Round 6 bidding guidance (http://www.communities.gov.uk/index.asp?id= 1128132#P80 _12915, accessed 20 September 2006).

CLG (2007a) New Projections of Households for England and the Regions to 2029 (http://comunities. gov.uk/index.asp?id=1002882&PressNoticeID=2374, accessed 6 July 2007).

CLG (2007b) Thames Gateway: North Kent (http://www.communities.gov.uk/index.asp?id= 1170138, accessed 17 July 2007).

CLG (2007c) *Planning for a Sustainable Future (Cm. 7120)* (HMSO, London).

Collins, A. (2004) 'Sexual dissidence, enterprise and assimilation: bedfellows in urban regeneration', *Urban Studies*, 41: 1789–1806.

Construction Task Force (1998) *Rethinking Construction* (HMSO, London).

Cook, I. (2004) *Waterfront Regeneration, Gentrification and the Entreprenuerial State: The Redevelopment of Gunwharf Quays*. Spatial Policy Analysis Working Paper 51 (School of Geography, University of Manchester, Manchester).

Cooper, I. (2000) 'Inadequate grounds for a "design-led" approach to urban renaissance? Towards an urban renaissance: final report of the urban task force', *Building Research and Information*, 28: 212–219.

Couch, C. and Denneman, A. (2000) 'Urban regeneration and sustainable development in Britain', *Cities*, 17: 137–147.

Coulter, J., Taylor, D., Whyte, S., Harloe, M., Glester, J., McGuire, J., Bounds, P., Gahagan, M. and Seviour, D. (2006) *Transition to transformation: Housing Market Renewal and our changing communities. A submission by the Market Renewal Pathfinder Chairs to the government's Comprehensive Spending Review 2007* (Bridging NewcastleGateshead, Newcastle).

Cowan, R. (2002) *Urban Design Guidance: Urban Design Frameworks, Development Briefs and Master Plans* (Thomas Telford, London).

CPRE Oxfordshire (2006) CPRE Oxfordshire Campaign Briefing: Planning Policy Statement 3 in Brief (http://www.cpreoxon.org.uk/news/briefing/edition/pps3.pdf, accessed 14 July 2007).

Crompton, J. (2001) 'Public subsidies to professional team sport facilities in the USA', in C. Gratton (ed.), *Sport in the City: The Role of Sport in Economic and Social Regeneration* (Routledge, London).

Dabinett, G. (2004) 'Creative Sheffield: creating value and changing values?', *Local Economy*, 19: 414–419.

Daly, G., Mooney, G., Poole, L. and Davis, H. (2005) 'Housing stock transfer in Birmingham and Glasgow: the contrasting experiences of two UK cities', *European Journal of Housing Policy*, 5: 327–341.

Davies, J. (2001) *Partnerships and Regimes: The Politics of Urban Regeneration in the UK* (Ashgate, Aldershot).

Davies, J. (2002) 'The governance of urban regeneration: a critique of the "governing without government" thesis', *Public Administration*, 80: 301–322.

Davoudi, S. (2000) 'Sustainability: a new "vision" for the British planning system', *Planning Perspectives*, 15: 123–137.

Deas, I. and Giordano, B. (2002) 'Locating the competitive city in England', in I. Begg (ed.), *Urban Competitiveness: Policies for Dynamic Cities* (Policy Press, Bristol).

Department for Transport (2003) *The Future of Air Transport (Cm. 6046)* (HMSO, London).

Department for Transport (2006) The Thames Gateway Transport Summary (http://www.dft.gov.uk/pgr/regional/strategy/growthareas/thamesgatewaytransportsummar6072, accessed 4 July 2007).

Department of Employment (1985a) *Employment: The Challenge for the Nation* (HMSO, London).

Department of Employment (1985b) *Lifting the Burden* (HMSO, London).

Department of Employment (1992) *People, Jobs and Opportunity* (HMSO, London).

Department of the Environment (1977) *Policy for the Inner Cities (Cmnd. 6845)* (HMSO, London).

Department of the Environment (1994) *Sustainable Development: the UK Strategy (Cm. 2426)* (HMSO, London).

DETR (1999) *A Better Quality of Life: strategy for sustainable development for the UK (Cm. 4345)* (TSO, London).

DETR (1999) *Towards an Urban Renaissance: The Report of the Urban Task Force chaired by Lord Rogers of Riverside* (Routledge, London).

DETR (2000a) *Our Towns and Cities: The Future – Delivering an Urban Renaissance* (HMSO, London).

DETR (2000b) *Planning Policy Guidance 3: Housing* (HMSO, London).

Diamond, J. (2004) 'Local regeneration initiatives and capacity building: whose "capacity" and "building" for what?', *Community Development Journal*, 39: 177–189.

Dicken, P. (2003) *Global Shift: Reshaping the Global Economic Map in the 21st Century* (SAGE, London).

Donovan, R., Evans, J., Bryson, J., Porter, L. and Hunt, D. (2005) *Large-scale Urban Regeneration and Sustainability: Reflections on the 'Barriers' Typology* (Centre for Environmental Research and Training Working Paper 05/01, University of Birmingham, Birmingham).

DTI (1998) *Our Competitive Future: Building the Knowledge-driven Economy* (HMSO, London).

DTI (2000) *Excellence and Opportunity: A Science and Innovation Policy for the 21st Century* (HMSO, London).

DTI (2001) *Opportunity for All in a World of Change* (HMSO, London).

DTI (2003a) *Competing in a Global Economy: The Innovation Challenge* (HMSO, London).

DTI (2003b) *A Modern Regional Policy for the United Kingdom* (HMSO, London).

DTI (2004) *A Practical Guide to Cluster Developments* (HMSO, London).

DTI (2006a) *Sustainable Construction Strategy Report* (Department of Trade and Industry, London).

DTI (2006b) *Construction Statistics Annual* (HMSO, London).

Duffy, H. (1995) *Competitive Cities: Succeeding in the Global Economy* (E & FN Spon, London).

Edinburgh City Council (2007) Waterfront Edinburgh: Granton Master Plan (http://www.edinburgh.gov.uk/CEC/Corporate_Services/Corporate_Communications/waterfrontintro/index.html, accessed 26 June 2007).

Edwards, A. (2007) 'It's the Liverpool Echo Arena: 10,600 seat arena takes our name', *Liverpool Echo*, 1 June: 2.

Edwards, M. (2008) 'Structures for development in Thames Gateway: getting them right' in Cohen, P. & Rustin, M. (eds.) *London's Turning: the prospect of Thames Gateway* (Ashgate, London).

Elkin, S. (1987) *City and Regime in the American Republic* (University of Chicago Press, Chicago).

English Partnerships (2007) Carbon Challenge (http://www.englishpartnerships.co.uk/carbonchallenge.htm, accessed 23 April 2007).

Evans, J. (2007) 'Wildlife corridors: an urban political ecology', *Local Environment*, 12: 129–152.

Evans, J. and Jones, P. (2007) 'Sustainable urban regeneration as a shared territory', *Environment and Planning A*, advance online publication.

Fearnley, R. (2000) 'Regenerating the inner city: lessons from the UK's City Challenge experience', *Social Policy and Administration*, 34: 567–583.

Ferguson, B. (2007) 'Trams still a long way down the line', *Evening News (Edinburgh)*, 2 July: 12.

Florida, R. (2002) *The Rise of the Creative Class and How It's Transforming Work, Leisure, Community and Everyday Life* (Basic Books, New York).

Florio, S. and Brownhill, S. (2000) 'Whatever happened to criticism? Interpreting the London Docklands Development Corporation's obituary', *CITY*, 4: 53–64.

Furbey, R. (1999) 'Urban 'regeneration': reflections on a metaphor', *Critical Social Policy*, 19: 419–445.

General Register Office for Scotland (2007) Household Projections for Scotland, 2004-based (http://www.gro-scotland.gov.uk/statistics/publications-and-data/household-projections-statistics/household-projections-for-scotland-2004-based/summary-of-the-results.html, accessed 6 July 2007).

Gilligan, A. (2007) 'Spin, hype and the truth about London's Olympic legacy', *Evening Standard*, 19 March: 19.

Ginsberg, N. (2005) 'The privatization of council housing', *Critical Social Policy*, 25: 115–135.

Glancey, J. (2004) 'In for a penny', *Society Guardian*, 29 April.

Gracey, H. (1973) 'The 1947 planning system: the plan-making process', in P. Hall, H. Gracey, R. Drewett and R. Thomas (eds), *The Containment of Urban England. Vol. 2: The Planning System* (George Allen and Unwin, London).

Gratton, C., Shibli, S. and Coleman, R. (2005) 'Sport and economic regeneration in cities', *Urban Studies*, 42: 985–999.

Greenhalgh, P. and Shaw, K. (2003) 'Regional development agencies and physical regeneration in England: can RDAs deliver the urban renaissance?', *Planning Practice and Research*, 18: 161–178.

Guy, S. and Shove, E. (2000) *A Sociology of Energy, Buildings and the Environment: Constructing Knowledge, Designing Practice* (Routledge, London).

Gwilliam, M., Bourne, C., Swain, C. and Prat, A. (1999) *Sustainable Renewal of Suburban Areas* (Joseph Rowntree Foundation, York).

Hall, S. and Nevin, B. (1999) 'Continuity and change: a review of English regeneration policy in the 1990s', *Regional Studies*, 33: 477–482.

Hammerton, F. (2005) 'Gateway to a £650m business boost and 16,500 new jobs', *Essex Chronicle*, 28 July: 8.

Harrison, C. and Davies, G. (2002) 'Conserving biodiversity that matters: practitioners' perspectives on brownfield development and urban nature conservation in London', *Journal of Environmental Management*, 65: 95–108.

Hart, T. and Johnston, I. (2000) 'Employment, education and training', in P. Roberts and H. Sykes (eds), *Urban Regeneration: A Handbook* (SAGE, London).

Harvey, D. (1989) 'From managerialism to entrepreneurialism: the transformation of urban governance in late capitalism', *Geografiska Annaler B*, 71: 3–17.

Healey, P. (1997) 'A strategic approach to urban regeneration', *Journal of Property Development*, 1: 105–110.

Henry, I. and Dulac, C. (2001) 'Sport and social regulation in the city: the cases of Grenoble and Sheffield', *Loisir et Société*, 24: 47–78.

Henry, I. and Paramio-Salcines, J. (1999) 'Sport and the analysis of symbolic regimes: a case study of the city of Sheffield', *Urban Affairs Review*, 34: 641–666.

HM Government (2005) *One Future – Different Paths: The UK's Shared Framework for Sustainable Development* (DEFRA, London).

HM Treasury (2001) *Productivity in the UK: The Regional Dimension* (HMSO, London).

HM Treasury (2005) *The Government's Response to Kate Barker's Review of Housing Supply* (HMSO, London).

Hughes, H. (2003) 'Marketing gay tourism in Manchester: new market for urban tourism or destruction of "gay space"?', *Journal of Vacation Marketing*, 9: 152–163.

Imrie, R. and Raco, M. (2003) 'Community and the changing nature of urban policy', in R. Imrie and M. Raco (eds), *Urban Renaissance? New Labour, Community and Urban Policy* (Policy Press, Bristol).

Imrie, R. and Thomas, H. (1993) *British Urban Policy and the Urban Development Corporations* (London, SAGE).

In Suburbia Partnership (2005) *In Suburbia: Delivering Sustainable Communities* (Civic Trust, London).

Jacobs, J. (1961) *The Death and Life of Great American Cities* (Random House, New York).

Jacobs, J. (1985) *Cities and the Wealth of Nations* (Random House, Toronto).

Jessop, B. (1994) 'Post-Fordism and the state', in A. Amin (ed.), *Post-Fordism: A Reader* (Blackwell, Oxford).

Jones, P. (2005) 'The suburban high flat in the post-war reconstruction of Birmingham, 1945–71', *Urban History*, 32: 323–341.

Jones, P. and Evans, J. (2006) 'Urban regeneration, governance and the state: exploring notions of distance and proximity', *Urban Studies*, 43: 1491–1509.

Jones, P. and Wilks-Heeg, S. (2004) 'Capitalising culture: Liverpool 2008', *Local Economy*, 19: 341–360.

Karadimitriou, N. (2005) 'Changing the way UK cities are built: the shifting urban policy and the adaptation of London's housebuilders', *Journal of Housing and the Built Environment*, 20: 271–286.

Kent Thameside Delivery Board (2005) Strategic Flood Risk Assessment of Kent Thameside (http://www.kt-s.co.uk/kts02/pdfs/FR_main.pdf, accessed 4 July 2007).

Krugman, P. (1996) 'Making sense of the competitiveness debate', *Oxford Review of Economic Policy*, 12: 17–25.

Laganside Corporation (2007) Bringing New Life to the River (http://www.laganside.com/about.asp, accessed 8 February 2007).

LDA (2007) London Development Agency (LDA) Opportunities Fund (http://www.lda.gov.uk/server/show/ConWebDoc.1231, accessed 14 July 2007).

Leather, P., Cole, I., Ferrari, E., Flint, J., Robinson, D., Simpson, C. and Hopley, M. (2007) *National Evaluation of the HMR Pathfinder Programme: Baseline Report* (CLG, London).

Lever, W. (2002) 'The knowledge base and the competitive city', in I. Begg (ed.), *Urban Competitiveness: Policies for Dynamic Cities* (Policy Press, Bristol).

Lightmoor (2007) Development Proposals (http://www.lightmoor.info/development_proposals.html, accessed 27 June 2007).

Lloyd, M. (2002) 'Urban regeneration and community development in Scotland: converging agendas for action', *Sustainable Development*, 10: 147–154.

Lynch, K. (1960) *The Image of the City* (Technology Press, London).

MacLeod, G. (2002) 'From urban entrepreneurialism to a "revanchist city"? On the spatial injustices of Glasgow's renaissance', *Antipode*, 34: 602–624.

Malpass, P. (1994) 'Policy making and local governance: how Bristol failed to secure City Challenge funding (twice)', *Policy and Politics*, 22: 301–312.

March, J. and Olsen, J. (1984) 'The new institutionalism: organizational factors in political life', *The American Political Science Review*, 74: 734–749.

McCarthy, J. (2006) 'The application of policy for cultural clustering: current practice in Scotland', *European Planning Studies*, 14: 397–408.

McLaughlin, M. (2007) 'Key players join in as £655m Clyde Gateway masterplan gets under way', *The (Glasgow) Herald*, 20 March: 3.

Mebratu, D. (1998) 'Sustainability and sustainable development: historical and conceptual review', *Environmental Impact Assessment Review*, 18: 493–520.

Mitchell, J. (1999) *Business Improvement Districts and Innovative Service Delivery* (PriceWaterhouseCoopers, Arlington, VA).

Montgomery, J. (2003) 'Cultural quarters as mechanisms for urban regeneration. Part 1: conceptualising cultural quarters', *Planning Practice and Research*, 18: 293–306.

Montgomery, J. (2005) 'Beware 'the Creative Class'. Creativity and Wealth Creation Revisited'. *Local Economy*, 20(4): 337–343.

Mooney, G. (2004) 'Cultural policy as urban transformation? Critical reflections on Glasgow, European City of Culture 1990', *Local Economy*, 19: 327–340.

Moore, S. and Rydin, Y. (2008) 'Promoting sustainable construction: European and British networks at the knowledge–policy interface', *Journal of Environmental Policy and Planning*, 10(2).

Nathan, M. and Unsworth, R. (2006) 'Beyond city living: remaking the inner suburbs', *Built Environment*, 32: 235–249.

National Audit Office (2007) *The Thames Gateway: Laying the Foundations* (HMSO, London).

National Statistics (2006) 2003-based National and Sub-national Household Projections for Wales (http://new.wales.gov.uk/legacy_en/keypubstatisticsforwales/content/publication/housing/2006/sdr30-2006/sdr30-2006.pdf, accessed 6 July 2007).

Neighbourhood Renewal Unit (2007) Neighbourhood Renewal Fund (http://www.neighbourhood.gov.uk/page.asp?id=611, accessed 23 April 2007).

New East Manchester (2001) *New Town in the City* (NEM, Manchester).

Newman, J. (2001) *Modernising Governance: New Labour, Policy and Society* (SAGE, London).

Newman, O. (1973) *Defensible Space: People and Design in the Violent City* (Architectural Press, London).

Noon, D., Smith-Canham, J. and Eagland, M. (2000) 'Economic regeneration and funding', in H. Sykes and P. Roberts (eds), *Urban Regeneration: A Handbook* (SAGE, London).

Northern Ireland Assembly (2002) *Measures of Deprivation: Noble versus Robinson*. Northern Ireland Assembly Research Paper 02/02 (Northern Ireland Assembly, Belfast).

Northern Ireland Statistics and Research Agency (2007) Household Projections for Northern Ireland: 2002–2025 (http://www.nisra.gov.uk/statistics/financeandpersonnel/DMB/publications/householdexec.pdf, accessed 7 July 2007).

ODPM (2002a) Living Places: Cleaner, Safer, Greener (ODPM, London).

ODPM (2002b) *Planning Policy Guidance 17: Planning for Open Space, Sport and Recreation* (ODPM, London).

ODPM (2003) *Sustainable Communities: Building for the Future* (HMSO, London).

ODPM (2004a) *Skills for Sustainable Communities.* (ODPM, London).

ODPM (2004b) *Urban Regeneration Companies: Policy Stocktake* (ODPM, London).

ODPM (2005a) *Creating Sustainable Communities: Delivering the Thames Gateway* (HMSO, London).

ODPM (2005b) *Planning Policy Statement 1: Delivering Sustainable Development* (ODPM, London).

ODPM (2005c) *Securing the Future: UK Government Sustainability Strategy* (HMSO, London).

ODPM (2005d) *Sustainable Communities: People, Places and Prosperity* (HMSO, London).

ODPM (2006) *State of the English Cities Report* (ODPM, London).

Office for National Statistics (2002) England and Wales Census (http://www.statistics.gov.uk/census/GetData/default.asp accessed 29 May 2007).

Office for National Statistics (2007) Regional Profile: London (http://www.statistics.gov.uk/cci/nugget.asp?id=1132, accessed 26 December 2007).

Olympic Delivery Authority (2006) Media Release: London 2012 can be the 'Regeneration Games' – David Higgins (http://main.london2012.com/en/news/press+room/releases/2006/November/2006-11-22-12-25.htm, accessed 4 July 2007).

Owens, S. (1994) 'Land, limits and sustainability – a conceptual framework and some dilemmas for the planning system', *Transactions of the Institute of British Geographers*, 19: 439–456.

Parliamentary Office of Science and Technology (1998) *A Brown and Pleasant Land: Household Growth and Brownfield Sites* (HMSO, London).

Pendlebury, J. (1999) 'The conservation of historic areas in the UK: a case study of "Grainger Town", Newcastle upon Tyne', *Cities*, 16: 423–433.

Pendlebury, J. (2002) 'Conservation and regeneration: complementary or conflicting processes? The case of Grainger Town, Newcastle upon Tyne', *Planning Practice and Research*, 17: 145–158.

Pierson, P. and Skocpol, T. (2002) 'Historical institutionalism in contemporary political science', in I. Katznelson and H. Miller (eds), *Political Science: State of the Discipline* (Norton, New York).

Plummer, P. and Taylor, M. (2003) 'Theory and praxis in economic geography: "enterprising" and local growth in a global economy', *Environment and Planning C: Government and Policy*, 21: 633–649.

Porter, M. (1990) *The Competitive Advantage of Nations* (The Free Press, New York).

Prospect Magazine (2006) 'Carbuncles: comment' (http://www.prospectmagazine.com/carbuncles/carb_comments.php?nomid=24, accessed 23 February 2007).

Raco, M. (2005) 'Sustainable development, rolled-out Neo-Liberalism and sustainable communities', *Antipode*, 37: 324–346.

Raco, M. and Henderson, S. (2005) 'From problem places to opportunity spaces: the practices of sustainable urban regeneration'. Paper presented to the Sustainable Urban Brownfield Regeneration: Integrated Management Conference (University of Birmingham, 1 March 2005).

Raco, M. and Henderson, S. (2006) 'Sustainable planning and the brownfield development process in the United Kingdom', *Local Environment*, 11: 499–513.

Ratcliffe, J., Williams, B. and Branagh, S. (1999) *Managing and Financing Urban Regeneration: A Preliminary Study on the Prospective Use of Business Improvement Districts and Tax Increment Finance Districts in Ireland* (Dublin Institute of Technology, Dublin).

Rhodes, J., Tyler, P. and Brennan, A. (2003) 'New developments in area-based initiatives in England: the experience of the Single Regeneration Budget', *Urban Studies*, 40: 1399–1426.

Rhodes, J., Tyler, P. and Brennan, A. (2005) 'Assessing the effect of area-based initiatives on local area outcomes: some thoughts based on the national evaluation of the Single Regeneration Budget in England', *Urban Studies*, 42: 1919–1946.

Rhodes, R. (1997) *Understanding Governance: Policy Networks, Governance, Reflexivity and Accountability* (Open University Press, Buckingham).

Roberts, P. (2000) 'The evolution, definition and purpose of urban regeneration', in P. Roberts and H. Sykes (eds), *Urban Regeneration in the UK* (SAGE, London).

Robson, B. (2002) 'Mancunian ways: the politics of regeneration', in J. Peck and K. Ward, *City of revolution: restructuring Manchester* (Manchester University Press, Manchester pp. 34–47).

Robson, B., Bradford, M., Deas, I., Hall, E., Harrison, E., Parkinson, M., Evans, R., Garside, P. and Robinson, F. (1994) *Assessing the Impact of Urban Policy* (HMSO, London).

Rotherham Metropolitan Borough Council (2007) *Rotherham Renaissance: Summary* (Rotherham Metropolitan Borough Council, Rotherham).

Royal Bank of Scotland (2006) *First-time Buyer Property Index: Expert Advice on the Top Locations to Invest In* (Royal Bank of Scotland, Edinburgh).

Royal Institute of Chartered Surveyors (2005) *Green Value: Green Buildings, Growing Assets* (RICS, London).

Rydin, Y., Holman, N., Hands, V. and Sommer, F. (2003) 'Incorporating sustainable development concerns into an urban regeneration project: how politics can defeat procedures', *Journal of Environmental Planning and Management*, 46: 545–561.

Samuel, M. (2005) 'Death by tarmac: the sorry fate of Hackney Marshes in pursuit of our Olympic dream', *The Times*, 28 September: 79.

Sassen, S. (1994) *Cities in a World Economy* (Pine Forge Press, Thousand Oaks, CA).

Saxenian, A. (1999) *Silicon Valley's New Immigrant Entrepreneurs* (Public Policy Institute of California, San Francisco).

Scottish Executive (2007) Development Department (http://www.scotland.gov.uk/About/Departments/DD, accessed 16 February 2007).

Selman, P. (2002) 'Multi-function landscape plans: a missing link in sustainability planning?', *Local Environment*, 7: 283–294.

Shirley, P. and Box, J. (1998) *Biodiversity, Brownfield Sites and Housing: Quality of Life Issues for People and Wildlife* (The Urban Wildlife Partnership, Newark).

Simmie, J., Sennett, J. and Wood, P. (2002) 'Innovation and clustering in the London metropolitan region', in I. Begg (ed.), *Urban Competitiveness: Policies for Dynamic Cities* (The Policy Press, Bristol).

Smith, N. (1996) *The New Urban Frontier: Gentrification and the Revanchist City* (Routledge, London).

Social Exclusion Unit (2001) *A New Commitment to Neighbourhood Renewal: National Strategy Action Plan* (Cabinet Office, London).

Solihull Metropolitan Borough Council (2007) Regenerating North Solihull (http://www.solihull.gov.uk/section.asp?catid=756&docid=1224, accessed 26 June 2007).

Sprigings, N. (2002) 'Delivering public services under the New Public Management: the case of public housing', *Public Money and Management*, 22: 11–17.

Stewart, J. (1994) 'Between Whitehall and town hall: the realignment of urban regeneration policy in England', *Policy and Politics*, 22: 133–145.

Stone, C. (1989) *Regime Politics: Governing Atlanta, 1946–1988* (University Press of Kansas, Lawrence, KS).

Sustainable Construction Task Group (2003) The UK Construction Industry: Progress Towards More Sustainable Construction 2000–2003 (http://www.bre.co.uk/filelibrary/rpts/Progress Report.pdf, accessed 26 December 2007).

Sustainable Development Commission (2003) *Mainstreaming sustainable regeneration – a call to action (Part 1)*, (Sustainable Development Commission, London).

Swyngedouw, E. (2004) 'Globalisation or 'glocalisation'? Networks, territories and rescaling', *Cambridge Review of International Affairs*, 17: 25–48.

Talen, E. (1999) 'Sense of community and neighbourhood form: an assessment of the social doctrine of new urbanism', *Urban Studies*, 36: 1361–1379.

Thames Gateway (2007) Thames Gateway Transformal Locations (http://www.thamesgateway.gov.uk/70_ThamesGatewayTransformalLocations.html?PHPSESSID=1803e0fdcd2177bf6556cfd4ed41482e, accessed 4 July 2007).

Thornley, A., Rydin, Y., Scanlon, K. and West, K. (2005) 'Business privilege and the strategic planning agenda of the Greater London Authority', *Urban Studies*, 42: 1947–1968.

Travers, T. (2002) 'Decentralization London-style: the GLA and London governance', *Regional Studies*, 36: 779–788.

Urban Forum (2004) *Out of the SRB, into the Pot: What are the Implications for the Voluntary and Community Sector?* (Urban Forum, London).

Urban Task Force (1999) *Towards an Urban Renaissance.* Final Report of the Urban Task Force chaired by Lord Rogers of Riverside (DETR, London).

Urban Task Force (2005) *Towards a Strong Urban Renaissance: An Independent Report by Members of the Urban Task Force chaired by Lord Rogers of Riverside* (http://image. guardian.co.uk/sys-files/Society/documents/2005/11/22/UTF_final_report. pdf, accessed 21 August 2007).

Walburn, D. (2005) 'Trends in entrepreneurship policy', *Local Economy*, 20: 90–92.

Ward, K. (2006) 'Policies in motion', urban management and state restructuring: the trans-local expansion of BIDs. *International Journal of Urban and Regional Research*, 30: 54–76.

Ward, S. (1998) *Selling Places: The Marketing and Promotion of Towns and Cities 1850–2000* (Spon, London).

Watt, N. and Revill, J. (2007) 'New eco-towns to ease house crisis: Chancellor promises 100,000 low-emission homes to trump Cameron's green credentials', *The Observer*, 13 May: 6.

WEFO (2007) Approved Projects (http://www.wefo.wales.gov.uk/default.asp?action=approved-projects&ID=86, accessed 14 July 2007).

While, A. (2006) 'Modernism vs urban renaissance: negotiating post-war heritage in English city centres', *Urban Studies*, 43: 2399–2419.

Whitehand, J. and Carr, C. (2001) *Twentieth-century Suburbs: A Morphological Approach* (Routledge, London).

William, H. (2006) 'Olympic Games: London faces "extremely tight" timetable for 2012', *The Independent*, 13 May: 64.

Williams, R. (2007) 'Gypsies lose high court battle over Olympic sites', *The Guardian*, 4 May: 7.

Wolch, J. (1990) *The Shadow State: Government and Voluntary Sector in Transition* (The Foundation Centre, New York).

World Commission on Environment and Development (1987) *Our Common Future.* The Brundtland Report (Oxford University Press, Oxford).

Zukin, S. (1982) *Loft Living: Culture and Capital in Urban Change* (Johns Hopkins University Press, Baltimore, MD).

Index